Lecture Notes in Mathematics

Edited by A. Dold and B. Eckmann

1085

Geertrui K. Immink

Asymptotics of Analytic Difference Equations

Springer-Verlag
Berlin Heidelberg New York Tokyo 1984

Author

Geertrui K. Immink
Mathematisch Instituut, Rijksuniversiteit Utrecht
3508 TA Utrecht, The Netherlands

AMS Subject Classification (1980): 39A

ISBN 3-540-13867-6 Springer-Verlag Berlin Heidelberg New York Tokyo
ISBN 0-387-13867-6 Springer-Verlag New York Heidelberg Berlin Tokyo

Printing and binding: Beltz Offsetdruck, Hemsbach / Bergstr.
2146 / 3140-543210

Math Sci sep.

PREFACE

The present monograph is concerned with classes of difference equations of the type

$$\Phi(s,y(s),y(s+1)) = 0, \tag{0.1}$$

where s is a complex variable and Φ and y are n-dimensional vector functions.

The functions Φ that are considered are characterized approximately by the following properties.

(i) Φ is holomorphic in a set $S \times U \times U$, where S is an open sector and U is a neighbourhood of a point $y_o \in \mathbb{C}^n$.

(ii) Φ is represented asymptotically by a series of the form

$$\hat{\Phi}(s,y,z) = \sum_{h=o}^{\infty} \varphi_h(y,z)s^{-h/p}, \qquad (p \in \mathbb{N})$$

as $s \to \infty$ in S, and this asymptotic expansion is uniformly valid on all sets $S' \times U \times U$, where S' is a closed subsector of S.

(iii) The equation (0.1) possesses a formal solution $f = \sum_{h=o}^{\infty} f_h s^{-h/p}$ such that $f_o = y_o$.

We have derived existence theorems for analytic solutions of (0.1) that are represented asymptotically by the given formal solution. The method we have used is closely related to that employed for example by Wasow in his book on differential equations ([40]) and, in a much improved version, by Malgrange in [21].

Recent results of Ramis on differential equations ([29],[30]) have led us to examine the existence of solutions belonging to certain Gevrey classes of holomorphic functions.

Solutions of certain nonlinear equations of the type (0.1) may be used to simplify linear systems of difference equations. (A similar technique is employed in the theory of differential equations. See for example [34]). One of the main purposes of this study was to give a complete analytic theory for the homogeneous linear system

$$y(s+1) = A(s)y(s), \tag{0.2}$$

where A is an $n \times n$ matrix function which is meromorphic at infinity. This involves establishing the existence of a number of sectors and a corresponding number of fundamental matrices of (0.2) such that

(i) the sectors cover a full neighbourhood of infinity,
(ii) each fundamental matrix is holomorphic in the corresponding sector and is represented asymptotically by a given (and fixed) formal fundamental matrix as s tends to infinity in this sector.
A general result concerning this problem is stated in Chapter III (theorem 18.18).

Chapter I contains a selection of known results from the formal theory of linear difference equations and from the theory of Gevrey classes of holomorphic functions and series. The last section deals with the existence of right inverses of linear difference operators on Banach spaces of holomorphic functions with different types of asymptotic behaviour. The propositions stated in this section are fundamental for the rest of this work. The proofs of these propositions, which are very lengthy and rather technical, constitute the whole of Chapter II.

Chapter III is concerned with nonlinear difference equations and with block-diagonalization and -triangularization problems. In §18 the results of this chapter are applied to the analytic reduction of a homogeneous linear system to a canonical form.

This research was carried out at the University of Groningen under the guidance of professor B.L.J. Braaksma, whom I would like to thank for his advice and encouragement. Thanks are also due to professor M. van der Put and professor J.P. Ramis for their interest in my work and their careful reading of the manuscript. Finally, I am indebted to Martha Groen who produced the typewritten version of the manuscript.

CONTENTS

CHAPTER I. *LINEAR DIFFERENCE EQUATIONS.*

§1. Introduction.

A linear difference equation is an equation of the form

$$\sum_{h=o}^{n} a_h(s)y(s+h\omega) = b(s),$$

where $a_1,..,a_n$ and b are given complex-valued functions of the complex variable s and the 'difference' ω is a complex number. If neither a_o nor a_n vanishes identically the equation is said to be of the n th order. By means of the substitutions $s = \omega t$ and $y(\omega t) = z(t)$, it can be converted into

$$\sum_{h=o}^{n} a_h(\omega t)z(t+h) = b(\omega t).$$

A single (scalar) n th order linear difference equation is known to be equivalent to a system of n first order equations. Such a system has several advantages over the single equation, in particular the simplicity of the notation. For this reason we choose the second approach and consider systems (or vector equations) of the form

$$y(s+1) - A(s)y(s) = f(s), \tag{1.1}$$

where A(s) is an n by n matrix and f(s) is an n-dimensional vector. For the moment we shall assume that A and f are meromorphic functions of s in a neighbourhood of infinity. It is the purpose of this thesis to prove the existence of analytic solutions of (1.1) with a prescribed asymptotic behaviour as $|s| \to \infty$.

We begin by discussing some general properties of this type of equation. Since our theory is inspired by the corresponding theory of analytic linear differential equations, we wish furthermore to point out similarities and differences between the two fields.

First consider the system of homogeneous linear difference equations

$$y(s+1) - A(s)y(s) = 0. \tag{1.2}$$

If Y is a solution of the corresponding matrix equation then, obviously the same is true of $Y\Pi$, where Π is a matrix whose coefficients are periodic functions of s with period 1. A matrix solution Y of (1.2) in some region of the complex plane is called a fundamental matrix of (1.2) in that region if $Y\Pi \equiv 0$ implies $\Pi \equiv 0$.

Analytic solutions of (1.2) or, more generally, of (1.1) fall into two groups, according to the region of the complex plane in which they are defined. To see this, let us suppose that A and f are holomorphic functions for $|s| > R \geq 0$. Let y_1 be a solution of (1.1), analytic in the sector $S_1 = \{s \in \mathbb{C}: \alpha_1 < \arg s < \beta_1, |s| > R\}$, where $0 < \alpha_1 < \pi < \beta_1 < 2\pi$. Obviously, y_1 can be continued analytically into the U shaped region $U_1 = \{s \in \mathbb{C}: |s| > R, \operatorname{Re} s < 0 \text{ or } |\operatorname{Im} s| > R\}$. On the other hand, if A is invertible, then (1.1) is equivalent to the equation

$$y(s-1) = A(s-1)^{-1}\{y(s) - f(s-1)\}. \tag{1.3}$$

Assume that A^{-1} is holomorphic for $|s| > R$ and let y_2 be a solution of (1.1) (and hence of (1.3)) which is analytic in the sector $S_2 = \{s \in \mathbb{C}: \alpha_2 < \arg s < \beta_2, |s| > R\}$, where $-\pi < \alpha_2 < 0 < \beta_2 < \pi$. Then, apparently, y_2 can be continued analytically into the region $U_2 = \{s \in \mathbb{C}: |s| > R, \operatorname{Re} s > 0$ or $|\operatorname{Im} s| > R\}$. In particular, two fundamental matrices Y_1 and Y_2 of (1.1) which are holomorphic in S_1 and S_2, respectively, can be continued analytically into U_1 and U_2, respectively. The resulting matrix functions $\tilde{Y}_1$ and $\tilde{Y}_2$ are connected in the following way,

$$\tilde{Y}_2 = \tilde{Y}_1 \Pi_{12},$$

where the coefficients of Π_{12} are periodic functions of s with period 1 and holomorphic in the region $|\operatorname{Im} s| > R$.

For the reasons outlined above, we shall distinguish between 'right sectors' and 'left sectors', i.e. sectors S with the property that $s \in S$ implies $s + 1 \in S$, or $s \in S$ implies $s - 1 \in S$, respectively. In order to obtain a complete knowledge of the solutions of difference equations of the form (1.1) (with A invertible), in both categories of sectors, it is clearly sufficient to study (1.1) in right sectors and equations of the form

$$y(s-1) - A(s)y(s) = f(s)$$

in left sectors.

One way of dealing with an analytic difference or differential equation in a neighbourhood of infinity, is by inserting a power series in $\frac{1}{s}$ for the unknown function, and equating corresponding coefficients on both sides. It is known that for a relatively large class of differential equations such power series converge. But even if they are divergent they give useful information about the asymptotic behaviour of solutions of the equation as $|s| \to \infty$. The theory that is concerned with the existence of such power series solutions, regardless of their convergence, is usually called the formal theory of difference or differential equations. The formal theories of both types of equations show a great number of similarities. Some known results from the formal theory of difference equations are discussed in §2. They will play an important role throughout the rest of this thesis.

However, in the case of difference equations, formal power series solutions rarely converge. Even worse, a solution of (1.1) that is analytic in a sector, in general can not be continued analytically along any path encircling the origin.

We shall illustrate this by a simple example. But first, let us consider the effect of a singularity of A, A^{-1} or f on the solutions of (1.1).

Suppose that A, A^{-1} and f are meromorphic functions on $\mathbb{C}$. Let y_1 be a holomorphic solution of (1.1) in a left sector. By analytic continuation of y_1 one obtains a meromorphic function. If s_o is a pole of A or f, then, in general, y_1 has poles in the points $s_o + h$, $h \in \mathbb{N}$. Similarly, it can be seen from (1.3) that the analytic continuation of a solution y_2 which is holomorphic in a right sector, in general will have poles in the points $s_1 - h$, $h \in \mathbb{N} \cup \{0\}$, whenever s_1 is a pole of f or a zero of A.

Now, let f be defined by the series

$$f(s) = \sum_{n=o}^{\infty} \frac{1}{s^{n!}}, \qquad |s| > 1,$$

and consider the equation

$$y(s+1) - s\,y(s) = f(s).$$

It has a solution y that can be represented by the following series

$$y(s) = -\sum_{h=o}^{\infty} \sum_{n=o}^{\infty} \frac{1}{s(s+1)\ldots(s+h)(s+h)^{n!}} \tag{1.4}$$

It is easily verified that y is analytic in the region

$$G = \bigcap_{h \in \mathbb{N} \cup \{0\}} \{s \in \mathbb{C} : |s+h| > 1\}.$$

The unit circle is a natural boundary for f (cf.[36] , §4.7). We shall now show that the set C defined by

$$C = \bigcup_{h \in \mathbb{N}} \{s \in ¢ : |s + h| = 1, |\mathrm{Re}(s + h)| \leq \tfrac{1}{2}\} \cup \{s \in ¢ : |s| = 1, \mathrm{Re}\, s > -\tfrac{1}{2}\},$$

is a natural boundary for y.

Let $m \in \mathbb{N} \cup \{0\}$, $p, q \in \mathbb{N}$, $p < \frac{1}{3}q$. Furthermore we assume that $p > \frac{1}{6}q$ if $m > 0$. We choose a point $s \in G$ such that $\arg(s + m) = 2\frac{p}{q}\pi$ and $|s + m| = r > 1$ and consider the behaviour of y(s) as $r \to 1$. To this end we split the sum representing y(s) into the following two parts,

$$y_1(s) = -\sum_{h \neq m} \sum_{n=0}^{\infty} \frac{1}{s(s+1)..(s+h)(s+h)^{n!}} - \sum_{n=0}^{q-1} \frac{1}{s(s+1)..(s+m)(s+m)^{n!}}$$

and

$$y_2(s) = -\frac{1}{s(s+1)..(s+m)} \sum_{n=q}^{\infty} \frac{1}{(s+m)^{n!}} .$$

Obviously, $y_1(s)$ tends to a finite limit as $r \to 1$. If $n \geq q$ then q is a divisor of n!, hence $(s + m)^{n!} = r^{n!}$, which shows that $|y_2(s)| \to \infty$ if $r \to 1$. As the set of all points s that can be written in the form $s = m + e^{2p\pi i/q}$ for certain values of m,p and q as defined above is a dense subset of C, it follows that y cannot be continued analytically across C.
A similar observation can be made with respect to the homogeneous vector equation

$$\begin{pmatrix} y_1(s+1) \\ y_2(s+1) \end{pmatrix} = \begin{pmatrix} 1 & 0 \\ f(s) & s \end{pmatrix} \begin{pmatrix} y_1(s) \\ y_2(s) \end{pmatrix} ,$$

which possesses a fundamental matrix of the form

$$\begin{pmatrix} 1 & 0 \\ y(s) & \Gamma(s) \end{pmatrix} ,$$

where y is the function defined in (1.4).

§2. Formal solutions and canonical forms.

Consider the system of homogeneous linear difference equations

$$y(s + 1) - A(s)y(s) = 0, \qquad (2.1)$$

where A(s) is an n x n matrix. We begin by discussing the scalar case. Let

$$A(s) = s^d \sum_{h=o}^{\infty} a_h s^{-h},$$

where $d \in \mathbb{Z}$, $a_h \in ¢$ for all $h \in \mathbb{N} \cup \{0\}$ and $a_o \neq 0$. If $d = 0$, $a_o = 1$ and $a_1 = 0$, then (2.1) is satisfied by a formal power series in s^{-1}. In general, it has a formal solution of the form

$$y(s) = \Gamma(s)^d \, a_o^s \, s^{a_1/a_o} \sum_{h=o}^{\infty} y_h s^{-h},$$

where $y_h \in ¢$ for all $h \in \mathbb{N} \cup \{0\}$, $y_o \neq 0$. Using Stirling's formula we get

$$y(s) = s^{ds} (a_o e^{-d})^s \, s^{a_1/a_o - d/2} \sum_{h=o}^{\infty} \tilde{y}_h s^{-h}.$$

As the convergence of the series $\sum_{h=o}^{\infty} a_h s^{-h}$ plays no role in the determination of formal solutions of (2.1), we may take A to be a formal Laurent series in s^{-1}, i.e. $A \in ¢[[s^{-1}]][s]$.

Now suppose that $n > 1$. Let $A \in G\ell(n; ¢[[s^{-1}]][s])$. If A has the form $A(s) = I + \sum_{h=2}^{\infty} A_h s^{-h}$, then again equation (2.1) is satisfied by a formal power series $y \in (¢[[s^{-1}]])^n$. In the general case, solutions of (2.1) can be obtained by first converting A by means of suitably chosen transformations (block diagonalizing and shearing transformations) into a matrix of a particularly simple form, the so called normal or canonical form of A. It is known (cf. [11], [28],[39]) that there exist a positive integer p and a matrix function $F \in G\ell(n; ¢[[s^{-1/p}]][s^{1/p}])$ such that the transformation

$$F(s+1)^{-1} A(s) F(s) \tag{2.2}$$

changes A into a block diagonal matrix

$$A_+^c(s) = \mathrm{diag}\{A_{+,1}^c(s), \ldots, A_{+,m}^c(s)\} \tag{2.3}$$

consisting of m blocks of the form

$$\left.\begin{aligned} &A_{+,j}^c(s) = (s+1)^{d_j(s+1)} s^{-d_j s} e^{q_j(s+1) - q_j(s)} \left(1 + \frac{1}{s}\right)^{G_j}, \\ &\text{where} \\ &d_j \in \frac{1}{p}\mathbb{Z}, \quad q_j(s) = \sum_{ph=o}^{p} \mu_{j,h} s^h, \quad \mu_{j,h} \in ¢, \end{aligned}\right\} \tag{2.4}$$

$G_j = \gamma_j I_{n_j} + N_j$, $\gamma_j \in ¢$, $0 \le \mathrm{Re}\gamma_j < \frac{1}{p}$ [1), N_j is a nilpotent $n_j \times n_j$ matrix; $j = 1,..,m$.)

We use the notation $\sum_{ph=o}^{p}$ to indicate that h runs through the set $\frac{1}{p}\{0,..,p\}$. Putting $\mathrm{diag}\{d_1 I_{n_1},..,d_m I_{n_m}\} = D$, $\mathrm{diag}\{q_1(s)I_{n_1},..,q_m(s)I_{n_m}\} = Q(s)$ and $\mathrm{diag}\{G_1,..,G_m\} = G$, we have

$$A_+^c(s) = (s+1)^{D(s+1)}\, s^{-Ds}\, e^{Q(s+1)-Q(s)}\left(1+\frac{1}{s}\right)^G. \tag{2.5}$$

We shall call A_+^c a canonical form of A. The corresponding system of homogeneous linear difference equations can be solved immediately. Thus we find that the original equation (2.1) possesses a formal fundamental matrix Y which is given by

$$Y(s) = F(s)s^{Ds}\, e^{Q(s)}\, s^G.$$

The matrices $A_{+,j}^c(s)$ defined in (2.2) may also be written in the following manner,

$$A_{+,j}^c(s) = s^{d_j}\{p_j(s)I_{n_j} + s^{-1}L_j + \tilde{A}_j(s)\},$$

where $p_j(s) = \sum_{ph=1}^{p} \lambda_{j,h}\, s^{h-1}$, $\lambda_{j,h} \in ¢$, $\lambda_{j,1} \neq 0$, $L_j = \lambda_{j,0}\, I_{n_j} + N_j$, $\lambda_{j,0} \in ¢$, $0 \le \mathrm{Re}(\lambda_{j,1}^{-1}\lambda_{j,0}) - \frac{1}{2}d_j < \frac{1}{p}$ and $\tilde{A}_j \in s^{-1-1/p}\,\mathrm{End}(n;¢[[s^{-1/p}]])$; $j = 1,..,m$.

More commonly in the theory of differential and difference equations, canonical forms are defined in such a way that $\tilde{A}_j \equiv 0$ for all $j \in \{1,..,m\}$. However, the exact form of the matrices $\tilde{A}_j(s)$ is of no real importance (It can be changed arbitrarily by means of a transformation of the type (2.2)). What interests us here is the values of the constants d_j and $\lambda_{j,h}$ $(j \in \{1,..,m\},\ h \in \frac{1}{p}\{0,..,p\})$. It can be proved that these are uniquely determined by A ([4],[10]). The following relations hold for all $j \in \{1,..,m\}$:

---

1) Note that the transformation $T(s+1)^{-1}A_+^c(s)T(s)$ with $T(s) = \mathrm{diag}\{I_{n_1},..,s^{r/p}I_{n_j},..,I_{n_m}\}$ has the effect of changing γ_j into $\gamma_j - \frac{r}{p}$. The choice of the interval $[0,\frac{1}{p})$ for $\mathrm{Re}\gamma_j$ is therefore arbitrary.

$$\mu_{j,1} = \log \lambda_{j,1} - d_j,$$

$$\mu_{j,h} = \frac{1}{h} \lambda_{j,h} \exp(-d_j - \mu_{j,1}) \quad \text{if } h \in \frac{1}{p}\{1,..,p-1\} \tag{2.6}$$

$$\gamma_j = \lambda_{j,o} \exp(-d_j - \mu_{j,1}) - \tfrac{1}{2} d_j$$

Hence we see that the numbers γ_j and, for $h \neq 1$, $\mu_{j,h}$, are uniquely determined by A, whereas the numbers $\mu_{j,1}$ are determined modulo $2\pi i$ by A. By $\underline{\mu_{j,1}}$ we shall denote the determination of $\mu_{j,1}$ such that $0 \leq \operatorname{Im} \underline{\mu_{j,1}} < 2\pi$, and by $\underline{q_j}$ the corresponding determination of the polynomial q_j.

In many problems it is not necessary to completely carry out the reduction of A to its canonical form.
Let $F_N(s)$ denote the sum of the first N terms of the series representing F(s). Since $F \in G\ell(n;\mathbb{C}[\![s^{-1/p}]\!][s^{1/p}])$, F_N will be invertible (and $F_N^{-1} \in G\ell(n;\mathbb{C}\{s^{-1/p}\}[s^{1/p}]))$ if N is sufficiently large. The transformed matrix $F_N(s+1)^{-1}A(s)F_N(s)$ may be written as follows

$$F_N(s+1)^{-1}A(s)F(s) = A_+^c(s) + s^{-r}\tilde{A}(s), \tag{2.7}$$

where $\tilde{A} \in \operatorname{End}(n;\mathbb{C}[\![s^{-1/p}]\!])$ and $r \in \mathbb{R}$. By taking N sufficiently large we can achieve that $r > 1 - \min_{j \in \{1,..,m\}} d_j$. In that case the canonical form A_+^c can be deduced from (2.7).

We have assumed so far that the coefficients of A are formal Laurent series in s^{-1}. More generally, if $A \in G\ell(n;\mathbb{C}[\![s^{-1/r}]\!][s^{1/r}])$ for some $r \in \mathbb{N}$, then A possesses a canonical form $A_+^c \in G\ell(n;\mathbb{C}[\![s^{-1/p}]\!][s^{1/p}])$, where p is a multiple of r, of the type described above. For convenience, we shall take p to be a fixed positive integer and consider the following set of matrices.

<u>DEFINITION</u>. *By* $\hat{M}_o$ *we denote the set of all* $A \in G\ell(n;\mathbb{C}[\![s^{-1/p}]\!][s^{1/p}])$ *with the property that* $A_+^c \in G\ell(n;\mathbb{C}[\![s^{-1/p}]\!][s^{1/p}])$.

The following sets of 'invariants' of the equation (2.1) (invariant that is, under transformations of the form (2.2)) will play an important role in the asymptotic theory discussed in this thesis.

DEFINITION. *Let* $A \in \hat{M}_o$ *and let* A_+^c *be a canonical form of* A *as defined in* (2.3) *and* (2.4). *We use the following notation:*

1. $d^+(A) = \{d_1, .., d_m\}$.
2. $\gamma^+(A) = \{\gamma_1, .., \gamma_m\}$.
3. $k^+(A)$ *is the set of all* $k \in \frac{1}{p}\{0, .., p\}$ *with the property that there is an integer* $j \in \{1, .., m\}$ *such that*

$$d_j = 0, \quad \text{degr } \underline{q}_j = k.$$ [2)]

4. *Let* $k \in k^+(A)$, $k \neq 0$. *Then* $\mu_k^+(A)$ *is the set of all* $\mu \in ¢$ *with the property that there is an integer* $j \in \{1, .., m\}$ *such that*

$$d_j = 0, \text{ degr } \underline{q}_j = k, \ \mu = \mu_{j,k}.$$

5. *Let* $j \in \{1, .., m\}$. *If* $d_j = 0$, $\mu_{1,j} \neq 0$, *then* $\Sigma_{1,j}^+(A)$ *is the set of all real numbers* α *such that*

$$\alpha = \frac{\pi}{2} - \arg \mu_{1,j}$$

(*or, equivalently,* $\Sigma_{1,j}^+(A) = \{\frac{\pi}{2} - \arg(\underline{\mu_{1,j}} + 2\ell\pi i), \ \ell \in \mathbb{Z}\}$). *Otherwise,* $\Sigma_{1,j}^+(A) = \phi$.

6. *Let* $k \in k^+(A)$, $k \neq 0$. $\Sigma_k^+(A)$ *is the set of all real numbers* α *with the property that*

$$\alpha = \frac{1}{k}\left(\frac{\pi}{2} - \arg \mu\right), \qquad \mu \in \mu_k^+(A).$$

The elements of $\Sigma_k^+(A)$ *will be called k-singular directions of* A.
Further, we define: $\Sigma_o^+(A) = \emptyset$ *and* $\Sigma^+(A) = \bigcup_{k \in k^+(A)} \Sigma_k^+(A)$. *In 5 and 6 all determinations of the arguments are allowed.*

The formal theory of the equation

$$y(s-1) - A(s)y(s) = 0 \tag{2.8}$$

is analogous to that of (2.1). If $A \in \hat{M}_o$ then there exists a matrix function $\tilde{F} \in G\ell(n; ¢[\![s^{-1/p}]\!][s^{1/p}])$ such that the transformation

$$\tilde{F}(s-1)^{-1}A(s)\tilde{F}(s)$$

2) Here we use the notation: $\text{degr} \sum_{ph=o}^{p} \mu_h s^h = \max\{h \in \frac{1}{p}\{0, .., p\} : \mu_h \neq 0\}$, degr $0 = 0$.

changes A(s) into a block diagonal matrix

$$\left.\begin{aligned} &A_-^c(s) = \mathrm{diag}\{A_{-,1}^c(s),\ldots,A_{-,m}^c(s)\}, \\ &\text{where} \\ &A_{-,j}^c(s) = (s-1)^{\tilde{d}_j(s-1)} s^{-\tilde{d}_j s} e^{\tilde{q}_j(s-1)-\tilde{q}_j(s)} (1-\tfrac{1}{s})^{\tilde{G}_j},\ j = 1,\ldots,m. \end{aligned}\right\} \quad (2.9)$$

$\tilde{d}_j, \tilde{q}_j$ and $\tilde{G}_j$ are defined in a similar way as d_j, q_j and G_j in (2.4). To distinguish between the canonical forms (2.5) and (2.9) we shall call A_+^c a right and A_-^c a left canonical form of A. From $A_-^c(s)$ we can derive sets of invariants for equation (2.8) in exactly the same manner as we did for (2.1). These sets will be denoted by $d^-(A)$, $\gamma^-(A)$, $k^-(A)$, etc. One can show that $d^-(A) = -d^+(A)$ and $k^-(A) = k^+(A)$. From now on we shall put

$$k^+(A) = k(A).$$

It is easily seen that $(A_+^c)^{-1}$ and $(A_-^c)^{-1}$ are a left and right canonical form of A^{-1}, respectively. Hence we deduce the relations

$$d^+(A^{-1}) = d^-(A), k(A^{-1}) = k(A), \mu_k^+(A^{-1}) = \mu_k^-(A) \quad \text{for all } k \in k(A), \text{ etc.}$$

§3. Asymptotics. Gevrey classes of series and functions.

In this section we introduce different classes of holomorphic functions with a specific asymptotic behaviour near infinity. Since we are not interested in the behaviour of such functions in any bounded region of the complex plane, we shall identify all functions that coincide for sufficiently large values of $|s|$. This is stated more precisely below.

Let F be a decreasing set of closed unbounded regions $\{G(R), R > 0\}$ of the complex plane such that $d(G(R),0) \to \infty$ as $R \to \infty$. By $B(F)$ we denote the set of complex-valued functions f with the properties that f is continuous on G(R) and holomorphic in int G(R) for some $R > 0$. We define an equivalence relation on $B(F)$ as follows: f and g are equivalent if there is a positive number R such that f and g are continuous on G(R), holomorphic in int G(R) and $f(s) = g(s)$ for all $s \in G(R)$.

The majority of holomorphic 'functions' occurring in this and the following sections have to be interpreted as equivalence classes defined with respect to a given set F.

DEFINITION. *Let* F *be a decreasing set of closed unbounded regions* $\{G(R), R>0\}$ *such that* $d(G(R),0)\to\infty$ *as* $R\to\infty$.

1. *By* $A_0(F)$ *we shall denote the set of all* $f\in B(F)$ *with the property that there exists a formal series* $\sum_{h=0}^{\infty} a_h s^{-h/p}$ *and a positive number* R *such that, for all* $N\in\mathbb{N}$,

$$\sup_{s\in G(R)} |s^{N/p}(f(s)-\sum_{h=0}^{N-1} a_h s^{-h/p})| < \infty.$$

The series $\sum_{h=0}^{\infty} a_h s^{-h/p}$ *is called the asymptotic expansion of* f. *We write:* $\hat{f}=\sum_{h=0}^{\infty} a_h s^{-h/p}$.

2. $A_{oo}(F)$ *is the set of all* $f\in A_o(F)$ *such that* $\hat{f}=0$ (i.e. *the series with coefficients zero).*

If $f\in A_o(F)$ and $\hat{f}=\sum_{h=0}^{\infty} a_h s^{-h/p}$ we shall put

$$f(s)-\sum_{h=0}^{N-1} a_h s^{-h/p} = R_N(f;s), \qquad N\in\mathbb{N}.$$

In general the quantity $\sup_{s\in G(R)} |s^{N/p} R_N(f;s)|$ will grow indefinitely as $N\to\infty$ (if not, f is holomorphic at infinity). For the following sets of functions, which were (re-)introduced and discussed in detail by Ramis (cf. [29],[30]) this growth rate is subject to certain restrictions.

DEFINITION. *Let* F *be a decreasing set of closed unbounded regions* $\{G(R), R>0\}$ *such that* $d(G(R),0)\to\infty$ *as* $R\to\infty$, *and let* $g>0$.

1. $A_g(F)$ *is the set of all* $f\in A_o(F)$ *with the property that there exist positive numbers* R, B *and* C *such that, for all* $N\in\mathbb{N}$,

$$\sup_{s\in G(R)} |s^{N/p} R_N(f;s)| \leq C(N!)^{1/pg} B^N.$$

We call $A_g(F)$ *a Gevrey class of holomorphic functions.*

2. *By* $A_{go}(F)$ *we denote the set of all* $f\in A_g(F)$ *such that* $\hat{f}=0$.

The coefficient a_h in the asymptotic expansion of a function $f\in A_g(F)$ ($g\geq 0$) is connected with the remainder $R_h(f;s)$ in the following way

$$a_h = \lim_{|s|\to\infty} s^{h/p} R_h(f;s).$$

DEFINITION. *Let* $g>0$. *By* $\mathbb{C}[[s^{-1/p}]]_{pg}$ *we denote the set of series* $\sum_{h=0}^{\infty} a_h s^{-h/p}$ *with the property that there exist positive numbers* B *and* C *such that, for all* $h \in \mathbb{N}$, *the following inequality holds*

$$|a_h| \leq C(h!)^{1/pg} B^h.$$

We call $\mathbb{C}[[s^{-1/p}]]_{pg}$ *a Gevrey class of formal power series in* $s^{-1/p}$.
Further, we define: $\mathbb{C}[[s^{-1/p}]]_o = \mathbb{C}[[s^{-1/p}]]$.
Obivously, $f \in A_g(F)$ implies $\hat{f} \in \mathbb{C}[[s^{-1/p}]]_{pg}$.

In the following sections we shall consider matrix functions with entries in $A_g(F)[s^{1/p}]$.

DEFINITION. *Let* $g \geq 0$.

(i) *By* $\hat{M}_g$ *we denote the set* $\hat{M}_o \cap G\ell(n;\mathbb{C}[[s^{-1/p}]]_{pg}[s^{1/p}])$.

(ii) *Let* F *be a decreasing set of closed unbounded regions* $\{G(R), R>0\}$ *such that* $d(G(R),0) \to \infty$ *as* $R \to \infty$. *By* $M_g(F)$ *we denote the set of matrix functions* $A \in G\ell(n;A_g(F)[s^{1/p}])$ *with the property that* $\hat{A} \in \hat{M}_g$.

If $g>0$, the elements of $A_{go}(F)$ are characterized by a property of exponential decrease at infinity.

PROPOSITION 3.1 (cf.[29]). *Let* F *be a decreasing set of closed unbounded regions* $\{G(R), R>0\}$ *such that* $d(G(R),0) \to \infty$ *as* $R \to \infty$, *and let* $g>0$. *Let* $f \in B(F)$. $f \in A_{go}(F)$ *if and only if there exist positive numbers* R *and* a *such that*

$$\sup_{s \in G(R)} e^{a|s|^g} |f(s)| < \infty.$$

In particular the above discussion applies to the case that F is a 'closed sector' in the sense we attribute to this notion in the following definition.

DEFINITION. *Let* $\alpha_1, \alpha_2 \in \mathbb{R}$, $0<\alpha_2-\alpha_1 \leq 2\pi$ *and let* $R>0$. *By* $S_R(\alpha_1,\alpha_2)$ *we denote the following subset of* $\mathbb{C}$

$$S_R(\alpha_1,\alpha_2) = \{s \in \mathbb{C} : \alpha_1 \leq \arg(s + Re^{i\alpha}) \leq \alpha_2 \text{ for all } \alpha \in [0,2\pi]\}$$

We define the 'closed sector' $S[\alpha_1,\alpha_2]$ *as follows*

$$S[\alpha_1,\alpha_2] = \{S_R(\alpha_1,\alpha_2);\ R>0\}.$$

The 'open sector' $S(\alpha_1,\alpha_2)$ *is defined by*

$$S(\alpha_1,\alpha_2) = \bigcup_{\alpha_1<\alpha_1'<\alpha_2'<\alpha_2} S[\alpha_1',\alpha_2'].$$

We shall write: $S[\alpha_1,\alpha_2] = \overline{S(\alpha_1,\alpha_2)}$. *Finally we define the 'half-open sectors'*

$$S[\alpha_1,\alpha_2) = \bigcup_{\alpha_1\le\alpha_1'<\alpha_2'<\alpha_2} S[\alpha_1',\alpha_2']$$

and

$$S(\alpha_1,\alpha_2] = \bigcup_{\alpha_1<\alpha_1'<\alpha_2'\le\alpha_2} S[\alpha_1',\alpha_2'] .$$

DEFINITION. *Let* S *be an open or half-open sector and let* $g\geq 0$. *The sets* $A_g(S)$ *and* $A_o(S)$ *are defined as follows*

$$A_g(S) = \bigcap_{\overline{S'}\subset S} A_g(\overline{S'})$$

and

$$A_{go}(S) = \bigcap_{\overline{S'}\subset S} A_{go}(\overline{S'}).$$

The following result is a generalization of the well-known Borel-Ritt theorem (cf. [30])

THEOREM 3.2. *Let* $g\geq 0$ *and let* $\sum_{h=o}^{\infty} a_h s^{-h/p} \in \mathbb{C}[\![s^{-1/p}]\!]_{pg}$. *If* $S = S(\alpha_1,\alpha_2)$ *is a sector with the property that* $g(\alpha_2-\alpha_1)\leq\pi$, *then there exists a function* $f\in A_g(S)$ *such that* $\hat{f} = \sum_{h=o}^{\infty} a_h s^{-h/p}$.

On the other hand, if $g>0$ and the sector S has an aperture greater than $\frac{\pi}{g}$, then an element of $A_g(S)$ is uniquely determined by its asymptotic expansion (cf. [22]):

THEOREM 3.3. *Let* $g>0$. *If* $S = S[\alpha_1,\alpha_2]$ *is a sector with the property that* $g(\alpha_2-\alpha_1)\geq\pi$, *then* $A_{go}(S) = \{0\}$.

§4. Right inverses of linear difference operators.

With a matrix function A we associate a right difference operator Δ_+ defined by

$$\Delta_+ y(s) = y(s+1) - A(s)y(s) \tag{4.1}$$

and a left difference operator Δ_-:

$$\Delta_- y(s) = y(s-1) - A(s)y(s). \tag{4.2}$$

For the sake of brevity we shall frequently use the notation $\pm$ or $\mp$, as in the following example: $\Delta_\pm y(s) = y(s\pm 1) - A(s)y(s)$. This expression contains both (4.1) and (4.2).

Let F be a decreasing set of closed unbounded regions $\{G(R),\ R>0\}$ with the property that $d(G(R),0) \to \infty$ as $R\to\infty$, such as a sector. If $A \in M_0(F)$ then $\hat{A} \in \hat{M}_0$. The invariants of the equation $y(s\pm 1) - \hat{A}(s)y(s) = 0$ (cf. §2) will be called formal invariants of the equation $\Delta_\pm y(s) = 0$. We shall write

$$d^\pm(\hat{A}) = d^\pm(A) = d(\Delta_\pm),$$

$$k\ (\hat{A}) = k\ (A) = k(\Delta_\pm),$$

$$\mu_k^\pm(\hat{A}) = \mu_k^\pm(A) = \mu_k(\Delta_\pm)\ \ (k \in k(\hat{A})), \text{ etc.}$$

Furthermore, the right and left canonical forms A_+^c and A_-^c of $\hat{A}$ will be called a right and left canonical form of A, respectively. The main purpose of this thesis is to establish conditions under which the matrix function A can be transformed into a canonical form $A_\pm^c$ by means of a transformation of the type

$$F(s\pm 1)^{-1}A(s)F(s),$$

with $F \in G\ell(n;A_0(F)[s^{1/p}])$. These conditions can be formulated in terms of the invariants defined on p. 8 . In particular, it will turn out that the singular directions are comparable to the Stokes lines in the theory of differential equations.

In this section we give sufficient conditions for the existence of right inverses of the operators Δ_+ and Δ_- on certain Banach spaces of holomorphic fucntions. In general we shall study equations of the type (4.1) in 'right sectors' and (4.2) in 'left sectors'.

DEFINITION. *Let* $\alpha_1,\alpha_2 \in \mathbb{R}$, $\alpha_1 < \alpha_2$. *The sector* $S(\alpha_1,\alpha_2)$, $S[\alpha_1,\alpha_2)$, $S(\alpha_1,\alpha_2]$ *or* $S[\alpha_1,\alpha_2]$ *is called a*

(i) *right sector if* $-\pi \leq \alpha_1 < 0 < \alpha_2 \leq \pi$

(ii) *strictly right sector if* $-\pi < \alpha_1 < 0 < \alpha_2 < \pi$

(iii) *left sector if* $0 \leq \alpha_1 < \pi < \alpha_2 \leq 2\pi$

(iv) *strictly left sector if* $0 < \alpha_1 < \pi < \alpha_2 < 2\pi$.

In addition to right and left sectors we shall consider sectors of the form $S[\alpha_1,\alpha_2)$ where $\alpha_1 = m\pi$, $m \in \mathbb{Z}$ and $0 < \alpha_2 - \alpha_1 \leq \pi$, or $S(\alpha_1,\alpha_2]$ where $\alpha_2 = m\pi$, $m \in \mathbb{Z}$ and $0 < \alpha_2 - \alpha_1 \leq \pi$. Furthermore, we shall frequently use sets of closed unbounded regions which are slightly more general than closed sectors:

DEFINITION *Let* $\alpha_1,\alpha_2 \in \mathbb{R}$, $0 < \alpha_2 - \alpha_1 \leq 2\pi$, *and let* $S = S[\alpha_1,\alpha_2]$. *A set of closed regions* $\{S(R), R > 0\}$ *will be called a right/left set of* S*-proper regions if for all* $R > 0$ *the following conditions are satisfied,*

(i) $\alpha_1 \leq \arg s \leq \alpha_2$ *and* $|s| \geq R$ *for all* $s \in S(R)$,

(ii) $s \in S(R)$ *implies* $s + 1 \in S(R)$/$s \in S(R)$ *implies* $s - 1 \in S(R)$,

(iii) *if* α_1' *and* α_2' *are real numbers such that* $S[\alpha_1',\alpha_2'] \subset S(\alpha_1,\alpha_2)$, *then there exists a positive number* R' *such that* $S_{R'}(\alpha_1',\alpha_2') \subset S(R)$.

If S is a right or left sector then a set of S-proper regions is automatically a right or left set of S-proper regions, respectively. If no confusion is possible we shall simply speak of a set of S-proper regions.

We consider the following sets of vector functions.

DEFINITION *Let* $g \geq 0$, $a \in \mathbb{R}$, $r \in \mathbb{R}$ *and let* G *be a closed region of the complex plane. By* $B_{g,a,r}(G)$ *we denote the Banach space of all* n*-dimensional vector functions* f *that are continuous on* G, *holomorphic in* int G *and, in addition, have the property that*

$$\|f\|^G_{g,a,r} \equiv \sup_{s \in G} |f(s)\, s^{-r} \exp(-a|s|^g)| < \infty\ .^{3)}$$

We shall define an ordering on the set $\{(g,a,r): g \geq 0,\ a \in \mathbb{R},\ r \in \mathbb{R}\}$ in such a way that, if G is a closed region with the property that $d(G,0) > 0$, then $(g_1,a_1,r_1) < (g_2,a_2,r_2)$ implies $B_{g_1,a_1,r_1}(G) \subset B_{g_2,a_2,r_2}(G)$. Thus, let $g_1,g_2 \geq 0$,

3) By $|.|$ we denote the usual vector or matrix norm.

a_1, a_2, r_1 and $r_2 \in \mathbb{R}$. We say that $(g_1,a_1) < (g_2,a_2)$ if any of the following conditions is fulfilled,

$$g_1 < g_2 \quad \text{and} \quad a_1, a_2 > 0,$$

$$g_1 > g_2 \quad \text{and} \quad a_1, a_2 < 0,$$

$$g_1 = g_2 \neq 0 \text{ and } a_1 < a_2,$$

$$a_1 \leq 0 \leq a_2 \text{ and } a_1 \neq a_2.$$

Furthermore we shall put

$$(g,a) = (0,0) = 0 \quad \text{if} \quad g = 0 \quad \text{or} \quad a = 0.$$

We say that $(g_1,a_1,r_1) < (g_2,a_2,r_2)$ when

$$(g_1,a_1) < (g_2,a_2)$$

or

$$(g_1,a_1) = (g_2,a_2) \quad \text{and} \quad r_1 < r_2.$$

If G is a subset of $\mathbb{C}$ we shall denote the set $\{s \in \mathbb{C} : -s \in G\}$ by $-G$. As we shall often have to consider multi-valued functions, we use the following convention concerning the determination of the argument. Suppose that G is a set with the property that $s \in G$ implies $s + 1 \in G$. Unless explicity stated otherwise, we choose the determination of args in such a way that $\arg s \in (-\pi,\pi)$ if $s \in G$ and $\arg s \in (0,2\pi)$ if $s \in -G$.

The following propositions are essential for the rest of this dissertation.

<u>PROPOSITION</u> 4.3. *Let* $A \in M_o(S)$, *where* $\pm S$ *is a strictly right closed sector*, $S = S[\alpha_1,\alpha_2]$. *Let* $k(A) \cup \{0,1\} = \{k_o,..,k_\ell\}$, *where* $\ell \in \mathbb{N}$ *and* $0 = k_o < k_1 < .. < k_\ell = 1$. *Let* $i \in \{1,..,\ell\}$. *Assume that*

(i) $A^{-1} \in \mathrm{End}(n;\mathcal{A}_o(S))$

(ii) *for all* $k \in k(A)$ *such that* $k \geq k_i$ *either of the following two conditions is satisfied*

$$\alpha_2 - \alpha_1 < \frac{\pi}{k}, \text{ and } \{\alpha_1, \alpha_2 - \frac{\pi}{k}\} \cap \Sigma_k^{\pm}(A) = \phi \tag{4.4}$$

or

$$\alpha_2 - \alpha_1 \geq \frac{\pi}{k} \text{ and } [\alpha_1, \alpha_2 - \frac{\pi}{k}] \cap \Sigma_k^{\pm}(A) = \phi \tag{4.5}$$

(iii) *if* $\alpha_2 - \alpha_1 < \pi$ *then* $[\alpha_2 - \pi, \alpha_1]$ *contains at most one element of* $\Sigma^{\pm}_{1,j}(A)$ *for each* $j \in \{1,..,m\}$ (cf. definition on p.8).

Then there exist a set of S-proper regions $\{S(R), R > 0\}$, *a real number* v_i, *nonnegative numbers* b_i, r_o *and a positive number* c_i *such that for all pairs* (g,a) *satisfying the condition*

$$(k_{i-1}, b_i) \leq (g, |a|) < (k_i, c_i) \tag{4.6}$$

there exist a positive constant R_o *and linear mappings*

$$\Lambda^{S(R)}_{g,a,r} : B_{g,a,r}(S(R)) \longrightarrow B_{g,a,r+v_i}(S(R))$$

which are defined for all $R \geq R_o$ *and for all* r *such that*

$$(g, |a|, |r|) > (0,0,r_o),$$

and possess the following properties:

1. $\Lambda^{S(R)}_{g,a,r}$ *is a right inverse of the difference operator* $\Delta_{\pm}$ *corresponding to* A.
2. *There exists a constant* K *independent of* R, *such that*

$$\|\Lambda^{S(R)}_{g,a,r} f\|^{S(R)}_{g,a,r+v_i} \leq K \|f\|^{S(R)}_{g,a,r}$$

for all $R \geq R_o$ *and all* $f \in B_{g,a,r}(S(R))$.

3. *If* g', a' *and* r' *are real numbers satisfying the conditions*

$$(k_{i-1}, b_i) \leq (g', |a'|) < (k_i, c_i)$$

and

$$(0,0,r_o) < (g',a',r') < (g,a,r) \text{ or } (g',a',r') < (g,a,r) < (0,0,-r_o),$$

then there exists a positive number R_1 *such that, for all* $R \geq R_1$,

$$\Lambda^{S(R)}_{g,a,r} \big|_{B_{g',a',r'}(S(R))} = \Lambda^{S(R)}_{g',a',r'} .$$

The proof of this proposition is given in Chapter II. It is based on an explicit construction of the regions $S(R)$ and the mappings $\Lambda^{S(R)}_{g,a,r}$. A close examination of this proof will reveal further details concerning the constants v_i, b_i, c_i and r_o. Thus, for instance, it can be seen that, if $0 \notin k(A)$, then r_o may be chosen equal to 0.

There is one more property of the mappings $\Lambda^{S(R)}_{g,a,r}$ that we shall need later on. Let $S' = S[\alpha_1',\alpha_2']$ be a closed left or right sector such that $S' \subset S(\alpha_1,\alpha_2)$. Suppose that for all $k \in k(A)$ such that $k \geq k_i$ the condition

$$[\alpha_2' - \frac{\pi}{k}, \alpha_2 - \frac{\pi}{k}) \cap \Sigma_k^{\pm}(A) = (\alpha_1,\alpha_1'] \cap \Sigma_k^{\pm}(A) = \phi \tag{4.7}$$

is fulfilled. Since, by assumption, S satisfies the conditions of proposition 4.3, the same is true of S'. Hence there exist S'-proper regions $S'(R)$ $(R>0)$ and linear mappings $\Lambda^{S'(R)}_{g,a,r}$ defined on $B_{g,a,r}(S'(R))$ for suitable values of g,a,r and R and possessing the properties mentioned in the proposition. Now the regions $S(R)$, $S'(R)$ and the mappings $\Lambda^{S(R)}_{g,a,r}$ and $\Lambda^{S'(R)}_{g,a,r}$ may be constructed in such a way that, for all $f \in B_{g,a,r}(S(R))$,

$$\Lambda^{S(R)}_{g,a,r} f|_{S'(R)} = \Lambda^{S'(R)}_{g,a,r} (f|_{S'(R)}) , \tag{4.8}$$

provided that g,a,r satisfy the conditions of proposition 4.3 with respect to both S and S' and R is a sufficiently large number.

Under far more restrictive conditions than those imposed in proposition 4.3, similar results can be obtained for left and right sectors which are not strictly left or right sectors.

<u>PROPOSITION</u> 4.9. *Let* $j \in \{1,2\}$ *and let* α_1,α_2 *be real numbers such that*

$$\alpha_j = m\pi,\ m \in \mathbb{Z},\ \textit{and}\ \pi < \alpha_2 - \alpha_1 < 2\pi.$$

Let $S = S[\alpha_1,\alpha_2]$ *and* $A \in M_o(S)$. *Using the notation of proposition* 4.3, *let* $i \in \{1,..,\ell\}$ *and assume that*

(i) $A^{-1} \in \mathrm{End}(n;A_o(S))$,

(ii) $0 \notin k(A)$ [4)],

(iii) *for all* $k \in k(A)$ *such that* $k \geq k_i$ *and for all* $\mu \in \mu_k^{\pm}(A)$ *the following condition is satisfied*

$$\mathrm{Re}(\mu e^{ik\alpha_j}) < 0.$$

(iv) *if there is a* $k \in k(A)$ *such that* $k \geq k_i$ *and* $\alpha_2 - \alpha_1 < \frac{\pi}{k}$ *then* $\alpha_2 - \frac{\pi}{k} \notin \Sigma_k^{\pm}(A)$.

(*We use the upper sign when* m *is odd, the lower sign when* m *is even*).
Then there exist a set of S-*proper regions* $\{S(R), R>0\}$, *a real number* v_i, *a nonnegative number* b_i *and a positive number* c_i *such that for all pairs* (g,a)

4) If $0 \in k(A)$ then the conclusions of the proposition remain valid for all pairs (g,a) satisfying, in addition to (4.10), the condition $(g,a) > 0$.

satisfying the condition

$$(k_{i-1},b_i) \leq (g,a) < (k_i,c_i), \tag{4.10}$$

and all $r \in \mathbb{R}$, *there exist a positive constant* R_o *and linear mappings*

$$\Lambda^{S(R)}_{g,a,r} : B_{g,a,r}(S(R)) \longrightarrow B_{g,a,r+v_i}(S(R))$$

which are defined for all $R \geq R_o$ *and possess the properties* 1,2 *and* 3 *mentioned in proposition* 4.3.*(The* 3 rd *property has to be modified in an obvious manner;* r_o *may be chosen equal to zero there.)*
In the case that $i = 1$ *the above statements are also true for all* g *and* a *such that*

$$(k_i, -c_i) < (g,a) \leq 0$$

and all $r \in \mathbb{R}$. (*Note that in this case* $g \leq \min k(A)$).

As in the previous case the regions S(R) and the right inverses $\Lambda^{S(R)}_{g,a,r}$ may be defined in such a way that the mappings $\Lambda^{S(R)}_{g,a,r}$ possess the property mentioned directly after proposition 4.3.

<u>PROPOSITION</u> 4.11. *Let* $\pm S = S[-\frac{\pi}{2},\frac{\pi}{2}]$ *and* $A \in M_o(S)$. *Assume that* $0 \notin d^{\pm}(A)$. *Then there exist a family of* S*-proper regions* $\{S(R), R > 0\}$, *a real number* v *and a positive number* c *such that, for all pairs* (g,a) *satisfying the condition*

$$0 \leq (g,|a|) < (1,c),$$

there exist a positive constant R_o *and linear mappings*

$$\Lambda^{S(R)}_{g,a,r} : B_{g,a,r}(S(R)) \longrightarrow B_{g,a,r+v}(S(R))$$

which are defined for all $R \geq R_o$ *and all* $r \in \mathbb{R}$ *and possess the following properties in addition to properties* 1 *and* 2 *of proposition* 4.3:

1^1. $\Lambda^{S(R)}_{g,a,r}$ *is a left inverse of the difference operator* $\Delta_{\pm}$ *associated with* A.

3^1. *If* $(g',a',r') < (g,a,r)$ *and if, in addition,*

$$(g', -a') < (1,c)$$

then there exists a positive number R_1 *such that, for all* $R > R_1$,

$$\Lambda^{S(R)}_{g,a,r}\,|_{B_{g',a',r'}(S(R))} = \Lambda^{S(R)}_{g',a',r'} .$$

Since our primary concern in this thesis is with solutions of difference equations that can be represented asymptotically by means of power series (i.e. elements of $A_o(S)$), we are mainly interested in right inverses defined on Banach spaces $B_{g,a,r}(S(R))$ with $a,r \leq 0$. With that restriction we can prove a considerably stronger result than the foregoing one.

PROPOSITION 4.12. *Let* $A \in M_o(S)$, *where* $\pm S$ *is a strictly right closed sector,* $S = S[\alpha_1,\alpha_2]$. *Using the notation of proposition* 4.3, *let* $i \in \{1,..,\ell\}$ *and assume that*

(i) $A^{-1} \in \mathrm{End}(n;A_o(S))$ *if* $\pm S \neq S[-\frac{\pi}{2},\frac{\pi}{2}]$

(ii) *for all* $k \in k(A)$ *such that* $k \geq k_i$ *the following condition is satisfied*

$$[\min\{\alpha_1,\alpha_2-\tfrac{\pi}{k}\},\ \max\{\alpha_1,\alpha_2-\tfrac{\pi}{k}\}] \cap \Sigma^{\pm}_k(A) = \phi \tag{4.13}$$

Then there exist a set of S*-proper regions* $\{S(R),R>0\}$, *a real number* v_i, *nonnegative numbers* b_i, r_o *and a positive number* c_i, *such that for all pairs* (g,a) *satisfying the condition*

$$(k_{i-1},b_i) \leq (g,-a) < (k_i,c_i), \tag{4.14}$$

there exist a positive constant R_o *and linear mappings*

$$\Lambda^{S(R)}_{g,a,r} : B_{g,a,r}(S(R)) \longrightarrow B_{g,a,r+v_i}(S(R))$$

which are defined for all $R \geq R_o$ *and for all* r *such that*

$$(g,a,r) < (0,0,-r_o), \tag{4.15}$$

and possess the properties 1,2 *and* 3 *of proposition* 4.3 *as well as property* 1^1 *of proposition* 4.11.

The observations made immediately after proposition 4.3 also apply in the case of proposition 4.12.

Finally, we shall consider sectors that are contained in upper or lower half planes.

PROPOSITION 4.16. *Let* $j \in \{1,2\}$ *and let* α_1, α_2 *be real numbers such that*

$$\alpha_j = m\pi,\ m \in \mathbb{Z},\ \text{and}\ 0 < \alpha_2 - \alpha_1 < \pi.$$

Let $S = S[\alpha_1, \alpha_2]$ *and* $A \in M_o(S)$. *Using the notation of proposition 4.3, let* $i \in \{1,..,\ell\}$ *and assume that*

(i) $A^{-1} \in \mathrm{End}(n; A_o(S))$,

(ii) *for all* $k \in k(A)$ *such that* $k \geq k_i$ *and all* $\mu \in \mu_k^{\pm}(A)$ *the following condition is satisfied*

$$\mathrm{Re}(\mu e^{ik\alpha_j}) > 0.$$

(*We use the upper sign when* m *is even, the lower sign when* m *is odd*).
Then there exist a set of S*-proper regions* $\{S(R), R > 0\}$, *a real number* v_i, *nonnegative numbers* b_i, r_o *and a positive number* c_i *such that for all* (g,a) *satisfying the condition* (4.14) *there exist a positive constant* R_o *and linear mappings* $\Lambda_{g,a,r}^{S(R)}$ *as in the previous proposition.*

REMARK 1. The condition $A \in M_o(S)$ in the propositions 4.3 - 4.16 may be replaced by the slightly weaker condition $A \in M_o(\mathfrak{S})$, where $\mathfrak{S}$ is the set of S-proper regions mentioned in the proposition that is considered. Naturally, assumption (i) of propositions 4.3, 4.9, 4.12 and 4.16 can be changed accordingly.

REMARK 2. By a slight modification of the proof of proposition 4.9, using estimates similar to those derived in part (i) of the proof of proposition 4.16, it can be shown that the constant R_o may be chosen independent of r. This fact is used in §15.

CHAPTER II. *EXISTENCE PROOFS FOR RIGHT INVERSES OF DIFFERENCE OPERATORS.*

§5. A transformation from left to right.

In this section it will be demonstrated how, by means of a simple transformation, the results stated in propositions 4.3 - 4.16 with regard to right sectors may be derived from corresponding statements concerning left sectors.

Let S be a subset of $\mathbb{C}$ with the property that $s \in S$ implies $s+1 \in S$. If φ is any mapping defined on S we define a mapping φ^* on $-S$ by

$$\varphi^*(s) = \varphi(e^{-i\pi}s)$$

Let A be an n by n matrix function on S and let Δ_+ be the corresponding right difference operator defined in (4.1). We define a left difference operator Δ_+^* as follows,

$$\Delta_+^* y(s) = y(s-1) - A^*(s)\, y(s), \qquad s \in -S.$$

The equation

$$\Delta_+ y(s) = f(s), \qquad s \in S,$$

is equivalent to

$$\Delta_+^* y^*(t) = f^*(t), \qquad t \in -S.$$

Suppose that Λ^* is a right inverse of Δ_+^* defined on a vector space containing f^*. Then the latter equation has a solution y such that

$$y^* = \Lambda^* f^*.$$

Hence it follows that

$$y(s) = \Lambda^* f^*(e^{i\pi}s), \qquad s \in S.$$

If A_+^c is a right canonical form of A, then there exists an $F \in G\ell(n;\mathbb{C}[\![s^{-1/p}]\!][s^{1/p}])$ such that

$$F(s+1)^{-1}A(s)F(s) = A_+^c(s), \qquad s \in S.$$

Putting $s = e^{-i\pi}t$ and $F(e^{-i\pi}t) = G(t)$, we find

$$G(t-1)^{-1}A(e^{-i\pi}t)G(t) = A_+^c(e^{-i\pi}t), \qquad t \in -S,$$

which shows that $(A_+^c)^*$ is a left canonical form of A^*.

The following lemma is easily deduced from the observations made above.

LEMMA 5.1. *Let* S *be a closed right sector or a sector of the type mentioned in proposition* 4.16 *with* m *even. Let* $A \in M_o(S)$ *and let* Δ_+ *denote the right difference operator defined in* (4.1).

(i) *Suppose that the conditions of one of the propositions* 4.3 - 4.16 *are fulfilled. Then the left difference operator* Δ_+^* *satisfies the conditions of the same proposition with respect to* -S.

(ii) *Let* $\{S(R), R > 0\}$ *be a set of* S*-proper regions. Assume that there exist linear mappings*

$$\Lambda_{g,a,r}^{-S(R)} : B_{g,a,r}(-S(R)) \longrightarrow B_{g,a,r+v}(-S(R)),$$

defined for suitable values of g,a,r,v, *and* R, *and possessing the properties mentioned in one of the propositions* 4.3 - 4.16. *Then the linear mappings* $\Lambda = \Lambda_{g,a,r}^{S(R)}$ *defined by*

$$(\Lambda f)(s) = \Lambda_{g,a,r}^{-S(R)} f^*(e^{i\pi}s),\ f \in B_{g,a,r}(S(R)), \qquad s \in S(R),$$

are right inverses of Δ_+^* *possessing analogous properties.*

§6. A canonical form of $\Lambda_{g,a,r}^{S(R)}$.

The problem of finding linear mappings Λ possessing the properties mentioned in any of the propositions 4.3 - 4.16 may be reduced to the simpler task of constructing such mappings in the case that the matrix function A is in canonical form. This is achieved by means of a preliminary transformation of A into a form 'sufficiently resembling' the canonical form, followed by a perturbation argument due to Malgrange (cf. [21]).

In view of lemma 5.1 we shall limit the discussion throughout the rest of this chapter to left sectors and sectors of the type occurring in proposition 4.16. From now on we shall assume that all difference operators considered are left difference operators. Therefore we drop the - signs in Δ_-, A_-^c, $d^-(A)$, etc.

Suppose that A^c is a canonical form of a matrix function A satisfying the conditions of one of the propositions 4.3 - 4.16 with regard to a given sector S. Let $\{S(R), R>0\}$ be a set of S-proper regions and let g,a,r,v, and R_o be suitable real numbers such that, for all $R \geq R_o$, there exist linear mappings

$$(\Lambda^c)^{S(R)}_{g,a,r} : B_{g,a,r}(S(R)) \longrightarrow B_{g,a,r+v}(S(R)),$$

with the properties mentioned in the proposition concerned.

Let N be an integer such that

$$N > v.$$

There exists a matrix function

$$F \in \mathrm{End}(n;\mathbb{C}[s^{-1/p}][s^{1/p}]) \cap G\ell(n;\mathbb{C}\{s^{-1/p}\}[s^{1/p}])$$

transforming A into a matrix function A^F of the form

$$A^F(s) = F(s-1)^{-1}A(s)F(s) = A^c(s) + s^{-N}\tilde{A}(s),$$

where

$$\tilde{A} \in \mathrm{End}(n;\mathcal{A}_o(S)).$$

(cf. §2).

Let $R_1 \geq R_o$ be chosen so large that $\tilde{A} \in \mathrm{End}(n;B_{o,o,o}(S(R_1)))$. For all $R \geq R_1$ we define a mapping $L = L^{S(R)}_{g,a,r}$ from $B_{g,a,r}(S(R))$ into itself, by the formula

$$Lf = s^{-N}\tilde{A}\Lambda^c f, \quad f \in B_{g,a,r}(S(R)), \tag{6.1}$$

where we have put

$$(\Lambda^c)^{S(R)}_{g,a,r} = \Lambda^c.$$

Since the norms of the mappings $(\Lambda^c)^{S(R)}_{g,a,r}$ are bounded by a quantity which is independent of R, while the matrix $\tilde{A}$ is a bounded function on $S(R_1)$, there exists a constant C independent of R (but depending on g,a and r) such that

$$||L||^{S(R)}_{g,a,r} \leq C\, R^{v-N}.$$

Thus, $||L||^{S(R)}_{g,a,r}$ will be smaller than one if $R \geq R_2$, where R_2 is a sufficiently large positive number (which, in general, will depend on g, a and r).

Hence we may define linear mappings

$$\tilde{\Lambda} = \tilde{\Lambda}^{S(R)}_{g,a,r} = (I + L^{S(R)}_{g,a,r})^{-1}, \quad R \geq R_2, \tag{6.2}$$

from $B_{g,a,r}(S(R))$ into itself, and we have

$$||\tilde{\Lambda}||^{S(R)}_{g,a,r} \leq (1 - C R_2^{\nu-N})^{-1} , \quad R \geq R_2. \tag{6.3}$$

Let Δ, Δ^F and Δ^c denote the left difference operators corresponding to A, A^F and A^c, respectively.

Putting

$$\Lambda_F = \Lambda^c \tilde{\Lambda} \tag{6.4}$$

and multiplying from the left by Δ^F we obtain the identity

$$\Delta^F \Lambda_F = (\Delta^c + s^{-N}\tilde{A})\ \Lambda^c (I + s^{-N} \tilde{A}\Lambda^c)^{-1}.$$

Since, by assumption, Λ^c is a right inverse of Δ^c we conclude that

$$\Delta^F \Lambda_F = I.$$

Now, let q be the smallest rational number such that $s^{-q}F(s)^{-1}$ is bounded on $S(R_1)$. One readily verifies that the linear mappings $\Lambda^{S(R)}_{g,a,r-q}$ defined by

$$\Lambda f(s) = F(s)\Lambda_F(F(s-1)^{-1}f)(s),\ f \in B_{g,a,r-q}(S(R)),\ R \geq R_2, \tag{6.5}$$

are right inverses of Δ the norms of which are bounded by a constant independent of R. Furthermore, if $(g',a',r') < (g,a,r)$ and if the restriction of $(\Lambda^c)^{S(R)}_{g,a,r}$ to $B_{g',a',r'}(S(R))$ coincides with $(\Lambda^c)^{S(R)}_{g',a',r'}$, then it is readily apparent from (6.1) - (6.5) that the restriction of $\Lambda^{S(R)}_{g,a,r-q}$ to $B_{g,a,r-q}(S(R))$ will coincide with $\Lambda^{S(R)}_{g',a',r'-q}$. Similarly, (4.8) follows easily from the corresponding property of the mappings $(\Lambda^c)^{S(R)}_{g,a,r}$.

Now suppose that Λ^c is a left inverse of Δ^c as well. Noting that

$$\Lambda_F \Delta^F = \Lambda^c (I + s^{-N}\tilde{A}\Lambda^c)^{-1}\ (\Delta^c + s^{-N}\tilde{A}),$$

and multiplying both sides from the right by $\Lambda^c\Delta^c$ we find

$$\Lambda_F \Delta^F = \Lambda^c (I + s^{-N}\tilde{A}\Lambda^c)^{-1}\ (I + s^{-N}\tilde{A}\Lambda^c)\Delta^c = I.$$

It immediately follows that the mapping Λ defined in (6.5) is an inverse of Δ in this case.

It remains to be proved that the constant R_2 can be chosen independent of r if r is smaller than some fixed number r_1.

Thus, let $r \leq r_1$ and consider the definition of $\tilde{\Lambda}^{S(R)}_{g,a,r}$ in (6.2) . In order to stress the fact that, a priori, R_2 depends on r, we shall write $R_{(r)}$ instead of

R_2. Let $r' < r$, $R \geq R_{(r)}$ and let $f \in B_{g,a,r'}(S(R))$. Evidently, $f \in B_{g,a,r}(S(R))$.
From (6.1) and (6.2) we deduce that

$$\tilde{\Lambda}^{S(R)}_{g,a,r} f = f - s^{-N}\tilde{A}\Lambda^c\tilde{\Lambda}^{S(R)}_{g,a,r} f. \tag{6.6}$$

The function on the right-hand side is seen to be an element of $B_{g,a,r'}(S(R))$ provided that

$$r + v - N \leq r'.$$

(Since $N > v$, $r + v - N < r$).

By assumption, there exists a constant K independent of R such that, for all $\varphi \in B_{g,a,r}(S(R))$,

$$\|\Lambda^c\varphi\|^{S(R)}_{g,a,r+v} \leq K\|\varphi\|^{S(R)}_{g,a,r}, \tag{6.7}$$

provided $R \geq R_{(r)}$.

Utilizing (6.3) and (6.7) and setting $\tilde{\Lambda}^{S(R)}_{g,a,r} = \tilde{\Lambda}$ we find that, under the conditions mentioned above,

$$\|s^{-N}\tilde{A}\Lambda^c\tilde{\Lambda} f\|^{S(R)}_{g,a,r'} \leq K_1\|f\|^{S(R)}_{g,a,r'},$$

where K_1 is a constant independent of R.

Hence, with (6.6) it follows that there exists a constant K_2, independent of R, such that, for all $f \in B_{g,a,r'}(S(R))$,

$$\|\tilde{\Lambda} f\|^{S(R)}_{g,a,r'} \leq K_2 \|f\|^{S(R)}_{g,a,r'},$$

provided that $r + v - N \leq r' < r$ and $R \geq R_{(r)}$.

As a consequence, it is found that, for all $R \geq R_{(r)}$, the restriction of $\Lambda^{S(R)}_{g,a,r-q}$ to $B_{g,a,r'-q}(S(R))$ is a right inverse of Δ possessing all the required properties. The proof is completed by a simple induction argument.

§7. Definition of Λ^c. Introductory remarks.

For the definition of a right inverse of a (left) difference operator Δ several formulas are available in the literature (cf. [4], [25]).

Let A be an $n \times n$ matrix function which is holomorphic in a region G with the property that $s \in G$ implies $s - 1 \in G$, and let Δ be the corresponding left difference operator. Suppose that $Y_o(s)$ is a fundamental matrix of the homogeneous equation $\Delta y = 0$. Let f be an n-dimensional vector function and consider the equation

$$\Delta y = f \tag{7.1}$$

Under certain conditions, including the convergence of infinite sums or integrals, a solution of this equation may be represented by one of the following expressions

$$y(s) = -Y_o(s) \sum_{h=o}^{\infty} Y_o(s-h-1)^{-1} f(s-h) \tag{7.2}$$

$$y(s) = -A(s)^{-1} f(s) - Y_o(s) \int_{C_+(s)} Y_o(\zeta-1)^{-1}\{1-e^{2\pi i(s-\zeta)}\}^{-1} f(\zeta)d\zeta, \tag{7.3}$$

$$y(s) = -A(s)^{-1} f(s) - Y_o(s) \int_{C_-(s)} Y_o(\zeta-1)^{-1}\{1-e^{-2\pi i(s-\zeta)}\}^{-1} f(\zeta)d\zeta. \tag{7.4}$$

In the last two formulas $C_+(s)$ and $C_-(s)$ are suitable contours enclosing the points s-1, s-2, etc., but not s.

In principle, the paths of integration $C_+(s)$ or $C_-(s)$ may be chosen so as to start from (or end in) a fixed point s_o. In that case, however, the vector function y will have singularities in the points s_o, s_o-1, etc.
In order to remove this defect we introduce yet another formula which, in a sense, is a combination of (7.3) and (7.4) ,

$$y(s) = -A(s)^{-1} f(s) - Y_o(s) \int_{C_1(s)} Y_o(\zeta-1)^{-1}\{1-e^{2\pi i(s-\zeta)}\}^{-1} f(\zeta)d\zeta +$$

$$- Y_o(s)\left[\int_{C_2(s)} Y_o(\zeta-1)^{-1}\{1-e^{-2\pi i(s-\zeta)}\}^{-1} f(\zeta)d\zeta + \int_{C(s)} Y_o(\zeta-1)^{-1} f(\zeta)d\zeta\right]^{5)} . \tag{7.5}$$

Here C(s) is a path from a fixed point (which may be finite or infinite) to a point on the segment between s and s-1, say s-½, while $C_1(s)$ and $C_2(s)$ are paths going from s-½ to infinity, in such a way that, along $C_1(s)$ the imaginary part of ζ increases whereas it decreases along $C_2(s)$.

It is not difficult, at least formally, to verify that the vector functions y defined by (7.2) - (7.5) satisfy the equation (7.1) . Assuming that the conditions alluded to above are satisfied, we may define a right inverse Λ of Δ by putting

$$\Lambda f = y.$$

Moreover, if Λ is defined by means of formula (7.2) , it turns out to be an actual inverse of Δ. The same is true in the case that formulas (7.3) or (7.4) are used, if the following additional condition is fulfilled,

5) This formula is a generalization of a formula used by Nörlund in [25] (cf. chapter 4, §1).

$$C_{\pm}(s-1) = C_{\pm}(s) - 1,$$

where the upper sign corresponds to formula (7.3), the lower sign to (7.4).

If A^c is a matrix of the form (2.9) and Λ^c is a right inverse of the corresponding (left) difference operator, then, obviously, Λ^c can be written

$$\Lambda^c = \operatorname{diag}\{\Lambda_1^c, .., \Lambda_m^c\}.$$

It suffices, therefore, to prove the existence of right inverses Λ_j^c of the difference operators Δ_j^c corresponding to the blocks A_j^c, possessing the required properties, for $j = 1,..,m$.
This will be done in the following sections. By the use of formulas (7.2) - (7.5) we shall define linear mappings Λ_j^c on appropriate Banach spaces $B_{g,a,r}(S(R))$. In view of the discussion above it seems unnecessary to give an explicit proof of the fact that the mappings so obtained are indeed right inverses of the difference operators Δ_j^c. It is easily verified, however, with the help of the estimates that will be supplied for the integrands in the case that formula (7.3), (7.4) or (7.5) is used, or for the terms of the infinite sum in case we use (7.2).

Unfortunately, we have not succeeded in constructing right inverses Λ_j^c defined in reference to the same set of regions $\{S(R), R > R_o\}$ for all $j \in \{1,..,m\}$ in the most general case, which is the reason we had to state the results in five separate propositions. This would not be of much consequence, if we were exclusively interested in solutions of linear difference equations. In that case, we might even content ourselves with a less general version of proposition 4.3, which is obtained by imposing the additional condition that k(A) contains only one element. Accordingly, the very complicated definition of the regions S(R) in §9.1 might be replaced by a definition similar to that used by Wasow (cf. [40], §14.2). With the help of the block-triangularization and diagonalizations theorems discussed in Chapter III we could still prove theorems 18.6 - 18.18.
In the nonlinear case however, the above mentioned difficulty in constructing right inverses Λ_j^c does affect the generality of the results.

§8. *Two preparatory lemmas.*

The following technical lemmas (in particular the second one) will be frequently used in the sequel.

LEMMA 8.1. *Let* $x_1 \in \mathbb{R}$, $x_2 \in (x_1,\infty]$. *Let* u,φ_1 *and* φ_2 *be differentiable real functions on* (x_1,x_2). *Assume that, for all* $x \in (x_1,x_2)$ *the following inequalities hold*

$$\varphi_1'(x) \geq \delta\ , \tag{8.2}$$

$$u(x) \geq 1\ , \tag{8.3}$$

$$|u'(x)| \leq C\ , \tag{8.4}$$

$$|\varphi_2(x)| \leq C\ u(x)^{1-\varepsilon}\ , \tag{8.5}$$

$$|\varphi_2'(x)| \leq C\ u(x)^{-\varepsilon}\ , \tag{8.6}$$

where δ, C *and* ε *are positive constants. Further, let* $m \in \mathbb{N}$ *and* $\varphi = \varphi_1 + \varphi_2$. *There exists a positive number* K *depending exclusively on* δ, C, ε *and* m, *such that*

$$\int_{x_1}^{x_2} \exp\{\varphi(x_1) - \varphi(x)\}\ \left|\ \log\ \left\{\frac{u(x_1)}{u(x)}\right\}\right|^m dx \leq K.$$

PROOF: It follows from (8.2) that

$$\varphi_1(x) - \varphi_1(x_1) \geq \delta(x - x_1), \quad x \in (x_1,x_2). \tag{8.7}$$

Similarly, we derive from (8.3) and (8.4) the inequalities

$$|u(x) - u(x_1)| \leq C(x - x_1), \quad x \in (x_1,x_2) \tag{8.8}$$

and

$$|\log u(x) - \log u(x_1)| \leq C(x - x_1), \quad x \in (x_1,x_2). \tag{8.9}$$

Let

$$x_3 = \min\{x_2, x_1 + \frac{u(x_1)}{2C}\}. \tag{8.10}$$

By (8.8)

$$u(x) \geq u(x_1) - C(x - x_1), \quad x \in (x_1,x_2),$$

hence

$$u(x) \geq \tfrac{1}{2}u(x_1)$$

for all $x \in (x_1, x_3)$.

Without loss of generality we may assume that $\varepsilon \leq 1$. Utilizing (8.6) we obtain the inequality

$$|\varphi_2(x) - \varphi_2(x_1)| \leq C\{\tfrac{1}{2}u(x_1)\}^{-\varepsilon}\ (x - x_1), \qquad x \in (x_1, x_3). \tag{8.11}$$

If

$$u(x_1) \geq 2\left(\frac{2C}{\delta}\right)^{1/\varepsilon},$$

then, with (8.7), it follows that

$$\varphi(x) - \varphi(x_1) \geq \frac{\delta}{2}\ (x - x_1), \qquad x \in (x_1, x_3).$$

And thus, putting

$$I(x) = \exp\{\varphi(x_1) - \varphi(x)\}\ \left|\log\{\frac{u(x_1)}{u(x)}\}\right|^m,$$

and using (8.9) we find

$$\int_{x_1}^{x_3} I(x)dx \leq \int_0^\infty \exp(-\frac{\delta}{2}\eta)(C\eta)^m d\eta.$$

If, on the other hand,

$$u(x_1) \leq 2\left(\frac{2C}{\delta}\right)^{1/\varepsilon},$$

then we deduce from (8.10) and (8.11) that

$$|\varphi_2(x) - \varphi_2(x_1)| \leq \left(\frac{2C}{\delta}\right)^{1/\varepsilon - 1},$$

for all $x \in (x_1, x_3)$.

With the help of (8.7) and (8.9) we now obtain

$$\int_{x_1}^{x_3} I(x)dx \leq \exp\{\left(\frac{2C}{\delta}\right)^{1/\varepsilon - 1}\} \int_0^\infty \exp(-\delta\eta)(C\eta)^m d\eta.$$

Thus, if $x_3 = x_2$, then the proof of the lemma is completed. Now suppose that $x_3 < x_2$.

(8.5) implies that

$$|\varphi_2(x) - \varphi_2(x_1)| \leq C\{u(x)^{1-\varepsilon} + u(x_1)^{1-\varepsilon}\}, \qquad x \in (x_1, x_2).$$

Since

$$u(x_1) \leq 2C(x - x_1), \qquad \text{if } x \in (x_3, x_2),$$

and consequently, by (8.8),

$$u(x) \leq 3C(x-x_1), \qquad x \in (x_3,x_2),$$

it follows that

$$|\varphi_2(x)-\varphi_2(x_1)| \leq C_1(x-x_1)^{1-\varepsilon}, \qquad x \in (x_3,x_2),$$

where

$$C_1 = (2^{1-\varepsilon}+3^{1-\varepsilon})C^{2-\varepsilon}.$$

Hence, utilizing (8.7) and (8.9) , we conclude that

$$\int_{x_3}^{x_2} I(x)dx \leq \int_0^\infty \exp(-\delta\eta + C_1\eta^{1-\varepsilon})(C\eta)^m \, d\eta.$$

∎

The next lemma is in fact a corollary to the preceding one.

LEMMA 8.12. *Let* S *be a region of the complex plane such that* $d(S,0) \geq 1$. *Let*

$$\Phi(\zeta) = \mu\zeta - a|\zeta|^\gamma + \nu \log|\zeta| + \Psi(\zeta), \qquad \zeta \in S,$$

where $\mu \in \mathbb{C}$, $a \in \mathbb{R}$, $\nu \in \mathbb{R}$, $\gamma \in (0,1]$ *and* Ψ *is an analytic function on* S *with the following properties*

$$|\Psi(\zeta)| \leq C|\zeta|^{1-\varepsilon}, \qquad \zeta \in S,$$

$$|\Psi'(\zeta)| \leq C|\zeta|^{-\varepsilon}, \qquad \zeta \in S,$$

where C *and* ε *are positive numbers.*
Let $s \in S$, $\alpha \in \mathbb{R}$, $x_o \in (0,\infty]$, *and suppose that the segment or half-line* ℓ *defined by*

$$\ell = \{\zeta \in \mathbb{C} : \zeta = s + e^{i\alpha} x \,,\; x \in (0,x_o)\}$$

is contained in S. *Assume that there exists a* $\delta > 0$ *such that one of the following conditions is satisfied*

case (i) $\gamma < 1$ *and* $\mathrm{Re}(\mu e^{i\alpha}) \geq \delta$.

case (ii) $\gamma = 1$, $|a| \leq \mathrm{Re}(\mu e^{i\alpha}) - \delta$.

case (iii) $\gamma = 1$, $a < 0$. *Furthermore, there is a* $\beta \in [0,\frac{\pi}{2}]$ *such that* $|\alpha - \arg s| \leq \beta$ *and* $a\cos\beta \leq \mathrm{Re}(\mu e^{i\alpha}) - \delta$.

case (iv) $\gamma = 1$, $a \geq 0$ *and* $x_o < \infty$. *Let* $s + e^{i\alpha}x_o = s_o$. *There is a* $\beta \in [0,\frac{\pi}{2}]$ *such that* $|\arg(s-s_o) - \arg s_o| \leq \beta$ *and* $a\cos\beta \geq \delta - \mathrm{Re}(\mu e^{i\alpha})$.

Further, let P *be a polynomial with positive coefficients. There exists a positive number* K *which is fully determined by the constants* $\delta, a, \gamma, \nu, C$ *and* ε *and by the polynomial* P, *such that*

$$\int_{\ell} |\exp\{\Phi(s) - \Phi(\zeta)\} P(\log|\tfrac{s}{\zeta}|) d\zeta| \leq K.$$

PROOF: We define a function u on $(0, x_o)$ by

$$u(x) = |s + e^{i\alpha}x| , \qquad x \in (0, x_o).$$

Obviously,

$$u(x) \geq 1, \qquad x \in (0, x_o)$$

A simple geometrical consideration shows that

$$u'(x) = \cos\{\alpha - \arg(s + e^{i\alpha}x)\} , \tag{8.13}$$

hence we see that, for all $x \in (0, x_o)$,

$$|u'(x)| \leq 1.$$

Suppose that $\gamma = 1$. Let us define functions φ_1 and φ_2 on $(0, x_o)$ as follows

$$\varphi_1(x) = \mathrm{Re}\{\mu(s + e^{i\alpha}x)\} - a|s + e^{i\alpha}x| , \qquad x \in (0, x_o),$$

and

$$\varphi_2(x) = \nu \log|s + e^{i\alpha}x| + \mathrm{Re}\, \Psi(s + e^{i\alpha}x) , \qquad x \in (0, x_o).$$

Utilizing (8.13) we find that

$$\varphi_1'(x) = \mathrm{Re}(\mu e^{i\alpha}) - a \cos\{\alpha - \arg(s + e^{i\alpha}x)\}, \qquad x \in (0, x_o).$$

Furthermore, it is readily verified that φ_2 and φ_2' satisfy inequalities of the form (8.5) and (8.6) , respectively (where the constant C will now depend on ν). In view of lemma 8.1 it suffices to prove that in each of the cases (ii)-(iv) the condition

$$\mathrm{Re}(\mu e^{i\alpha}) - a \cos\{\alpha - \arg(s + e^{i\alpha}x)\} \geq \delta \tag{8.14}$$

is satisfied. In case (ii) this is obvious.
In case (iii) it immediately follows by observing that, for all $x \in (0, x_o)$,

$$|\alpha - \arg(s + e^{i\alpha}x)| \leq |\alpha - \arg s| \leq \beta ,$$

and hence

$$\cos\{\alpha - \arg(s + e^{i\alpha}x)\} \geq \cos \beta.$$

On the other hand, in case (iv) we have

$$|\alpha - \arg(s + e^{i\alpha}x) - \pi| \leq |\arg(s - s_o) - \arg s_o| \leq \beta,$$

for all $x \in (0,x_o)$. Consequently,

$$-\cos\{\alpha - \arg(s + e^{i\alpha}x)\} \geq \cos\beta ,$$

and (8.14) is seen to be valid in this case as well. The proof of case (i) is even directer and is therefore omitted.

§9. Proof of proposition 4.3.

§9.1. Introduction.

In view of the preceding discussion it suffices to study those cases where A is in canonical form and S is a strictly left sector.

In §9.2 we define a suitable set of S-proper regions $\{S(R), R > 0\}$. Although we start from an arbitrary matrix function $A \in M(S)$ it will soon become clear that the shape of the regions S(R) is determined by the canonical form A^c of A.

We then proceed, in §9.3 and §9.4 , to define right inverses of the difference operators corresponding to the individual blocks of A^c, i.e., we consider matrix functions B of the form

$$B(s) = (s-1)^{d(s-1)} s^{-ds} e^{q(s-1)-q(s)} (1 - \frac{1}{s})^G, \tag{9.1}$$

where $d \leq 0$, q is a polynomial in $s^{1/p}$ and G a constant matrix, and prove that, in all cases that are consistent with the conditions of proposition 4.3, there exist linear mappings $\Lambda^{S(R)}_{g,a,r}$ possessing the required properties. For notational convenience we shall assume that all 'blocks' B are of dimension n.

§9.2. Definition of the regions S(R).

The basic ideas of the construction that follows are simple and easily understood. However, the detailed elaboration of these involves a great deal of tedious calculations.

Let $S = S[\alpha_1,\alpha_2]$ be a strictly left sector and $A \in M(S)$. If R is a positive number, $S^*(R)$ will denote the closed region consisting of all points lying to the left of or on the curve σ^* with parameter representation

$$\sigma^*(\xi) = Re^{i\alpha_1}(\xi + i)^{1/\kappa}, \qquad \xi \in (-\infty,\infty), \tag{9.2}$$

where κ is the positive number defined by

$$\alpha_2 - \alpha_1 = \frac{\pi}{\kappa},$$

and the determination of $(\xi + i)^{1/\kappa}$ is taken so that $\arg(\xi + i)^{1/k} \in (0, \frac{\pi}{\kappa})$. It is readily verified that $S^*(R) \subset S(\alpha_1,\alpha_2)$ and, furthermore, that $\arg \sigma^*(\xi)$ is a monotonic function of ξ, decreasing from α_2 to α_1.

In essence the regions $S^*(R)$ already have the required form. The modifications that will be introduced below can be made arbitrarily small.

For all $\lambda > 0$ we define a mapping χ_λ (from S into the Riemann surface of logs) by

$$\chi_\lambda(s) = s^\lambda, \qquad s \in S.$$

(Note that $\chi_\kappa(S^*(R))$ is a half plane.)

Let

$$\alpha_\lambda^*(\xi) = \arg(\chi_\lambda \circ \sigma^*)'(\xi), \qquad \xi \in (-\infty,\infty).$$

A short calculation shows that

$$\alpha_\lambda^*(\xi) = (\lambda - \kappa) \arg \sigma^*(\xi) + \kappa\alpha_1, \qquad \xi \in (-\infty,\infty). \tag{9.3}$$

If $\lambda \leq \kappa$ the expression on the right-hand side represents a non-decreasing function of $\arg \sigma^*(\xi)$, hence the set $\chi_\lambda(S^*(R))$ is convex in that case.

By $k_1(A)$ and $k_2(A)$ we shall denote the sets

$$k_1(A) = \{k \in k(A)\ ,\ k < \kappa\},$$

and

$$k_2(A) = \{k \in k(A)\ ,\ k \geq \kappa\},$$

respectively. Let $N \in \mathbb{N}$ and let $\{\xi_0,\xi_1,..,\xi_N\}$ be a strictly increasing sequence of real numbers such that the following condition holds

$$\begin{aligned} &\text{if } \frac{1}{k}\,\alpha_k^*(\xi) \in \Sigma_k(A) \text{ for some } k \in k_1(A) \text{ and } \xi \in (-\infty,\infty) \\ &\text{then } \xi \in \{\xi_1,..,\xi_{N-1}\}. \end{aligned} \tag{9.4}$$

We shall put

$$\sigma^*(\xi_h) = s_h, \qquad h \in \{0,..,N\}.$$

Furthermore, we choose a number ρ such that

$$\max_{k \in k_1(A)} k \leq \rho < \kappa. \tag{9.5}$$

Then we know that $\chi_\rho(S^*(R))$ is a convex set, hence the segments ℓ_h defined by

$$\ell_h(\xi) = s_h^\rho + (s_{h+1}^\rho - s_h^\rho)\frac{\xi-\xi_h}{\xi_{h+1}-\xi_h}, \quad \xi \in [\xi_h,\xi_{h+1}], \quad h \in \{0,..,N-1\},$$

lie in $\chi_\rho(S^*(R))$. Consequently, the curve σ with parameter representation

$$\sigma(\xi) = \sigma^*(\xi) \qquad \text{if } \xi < \xi_o \text{ or } \xi > \xi_N ,$$

$$\sigma(\xi) = \{s_h^\rho + (s_{h+1}^\rho - s_h^\rho)\frac{\xi-\xi_h}{\xi_{h+1}-\xi_h}\}^{1/\rho} \qquad \text{if } \xi \in [\xi_h,\xi_{h+1}], \quad h \in \{0,..,N-1\},$$

is contained in $S^*(R)$.

It is easily seen that arg $\sigma(\xi)$ is a strictly decreasing function of ξ. This implies that

$$\text{every half-line from 0 to infinity intersects } \sigma \text{ in at most one point} \tag{9.6}$$

The closed region consisting of all points lying to the left of or on the curve σ will be denoted by $S_\rho(R)$. If α_1' and α_2' are real numbers such that $[\alpha_1',\alpha_2'] \subset (\alpha_1,\alpha_2)$, then it follows from (9.6) that $S_{R'}(\alpha_1',\alpha_2') \subset S_\rho(R)$, for all sufficiently large numbers R'. For the purpose of proving proposition 4.3 we can take $S(R) = S_\rho(R)$, provided ρ is suitably chosen (cf. (9.5), (9.21), (9.24) and (9.25)). The proof that (4.8) holds requires a slightly more sophisticated definition of the regions S(R), which will be given at the end of this section. First we shall collect some relevant information concerning the shape of the regions $\chi_\lambda(S_\rho(R))$ for $\lambda > 0$ (and, more particularly, for $\lambda \in (0,1]$) and see what further conditions have to be imposed on ρ. From now on we drop the subscript ρ in $S_\rho(R)$.

By $\alpha_\lambda^-(\xi)$ and $\alpha_\lambda^+(\xi)$ we denote the arguments of the left and right derivative of $\chi_\lambda \circ \sigma$ in ξ, respectively. If both derivatives are equal we set $\alpha_\lambda^-(\xi) = \alpha_\lambda^+(\xi) = \alpha_\lambda(\xi)$. Computation of these quantities for different parts of the curve σ yields

$$\begin{aligned}
\alpha_\lambda(\xi) &= \alpha_\lambda^*(\xi) = (\lambda-\kappa)\arg\sigma(\xi) + \kappa\alpha_1 && \text{if } \xi<\xi_o \text{ or } \xi>\xi_N,\\
\alpha_\lambda^-(\xi) &= (\lambda-\rho)\arg\sigma(\xi) + \arg(s_{h+1}^\rho - s_h^\rho) && \text{if } \xi\in(\xi_h,\xi_{h+1}],\ h\in\{0,..,N-1\},\\
\alpha_\lambda^+(\xi) &= (\lambda-\rho)\arg\sigma(\xi) + \arg(s_{h+1}^\rho - s_h^\rho) && \text{if } \xi\in[\xi_h,\xi_{h+1}),\ h\in\{0,..,N-1\}.
\end{aligned} \tag{9.7}$$

Furthermore, we define

$$\alpha_\lambda^- = \inf_{\xi\in(-\infty,\infty)} \alpha_\lambda^-(\xi) ,$$

and

$$\alpha_\lambda^+ = \sup_{\xi \in (-\infty,\infty)} \alpha_\lambda^+(\xi).$$

Let us consider the differences

$$\varepsilon_h^- = \alpha_\lambda^-(\xi_h) - \alpha_\lambda^*(\xi_h) \ , \quad h \in \{1,..,N-1\} \ , \tag{9.8}$$

and

$$\varepsilon_h^+ = \alpha_\lambda^+(\xi_h) - \alpha_\lambda^*(\xi_h) \ , \quad h \in \{1,..,N-1\} \ . \tag{9.9}$$

From (9.3) and (9.7) we deduce that

$$\varepsilon_h^- = \arg \{\frac{(s_h^{-1} s_{h-1})^\rho - 1}{(s_h^{-1} s_{h-1})^\kappa - 1}\} \ , \qquad h \in \{1,..,N-1\} \ ,$$

and

$$\varepsilon_h^+ = \arg \{\frac{(s_{h+1} s_h^{-1})^\rho - 1}{(s_{h+1} s_h^{-1})^\kappa - 1}\}, \qquad h \in \{1,..,N-1\} \ .$$

Putting $(s_{h+1} s_h^{-1})^\kappa = w$, we may write

$$\varepsilon_h^+ = \arg \left(\frac{w^{\rho/\kappa} - 1}{w-1}\right) = \frac{\rho}{\kappa} \arg \left[\int_0^1 \{1 + (w-1)u\}^{\rho/\kappa - 1} du\right].$$

It is easily seen that $\arg w \in (-\pi,0)$ and, consequently, $\arg w < \arg\{1 + (w-1)u\} < 0$ if $u \in (0,1)$. Making use of the fact that $\rho < \kappa$, we obtain

$$0 < \varepsilon_h^+ < (1 - \frac{\rho}{\kappa})\pi \quad \text{for all } h \in \{1,..,N-1\}. \tag{9.10}$$

A similar argument shows that

$$-(1 - \frac{\rho}{\kappa})\pi < \varepsilon_h^- < 0 \quad \text{for all } h \in \{1,..,N-1\}. \tag{9.11}$$

Now let us suppose that $\lambda \leq \rho$. In that case $\alpha_\lambda^+(\xi)$ is seen to be a monotonic, non-increasing function of $\arg \sigma(\xi)$, hence $\chi_\lambda(S(R))$ is a convex set (cf.fig.1).

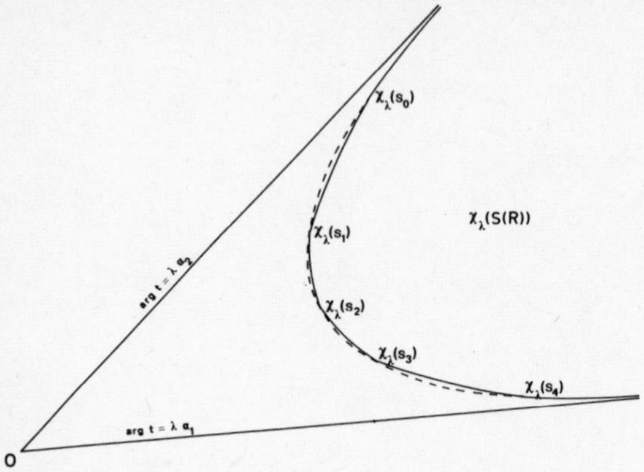

Fig.1.

Sketch of $\chi_\lambda(S(R))$ for $\lambda < \rho$, $N = 4$.

—: the curve $\chi_\lambda \circ \sigma$

--: the curve $\chi_\lambda \circ \sigma^*$

Moreover, we have

$$\alpha_\lambda^+ = \lim_{\xi \to +\infty} \alpha_\lambda(\xi) = \lim_{\xi \to +\infty} \alpha_\lambda^*(\xi) = \lambda\alpha_1 \,, \tag{9.12}$$

and

$$\alpha_\lambda^- = \lim_{\xi \to -\infty} \alpha_\lambda(\xi) = \lim_{\xi \to -\infty} \alpha_\lambda^*(\xi) = \lambda\alpha_2 - \pi. \tag{9.13}$$

In particular, we see that

$$\alpha_\lambda^- < \alpha_\lambda^+ < \alpha_\lambda^- + \pi \tag{9.14}$$

From the definition of α_λ^+ and α_λ^- we now deduce that

for any $t \in \chi_\lambda(S(R))$ and any $\alpha \in [\alpha_\lambda^+, \alpha_\lambda^- + \pi]$, the half-line ℓ from t to infinity with directional angle α is contained in $\chi_\lambda(S(R))$. (9.15)

Further, let $t_h = \chi_\lambda(s_h)$ for some $h \in \{1,..,N-1\}$. Then, for all $t \in \chi_\lambda(S(R))$, the following inequality holds:

$$\alpha_\lambda^+(\xi_h) \le \arg(t - t_h) \le \alpha_\lambda^-(\xi_h) + \pi.$$

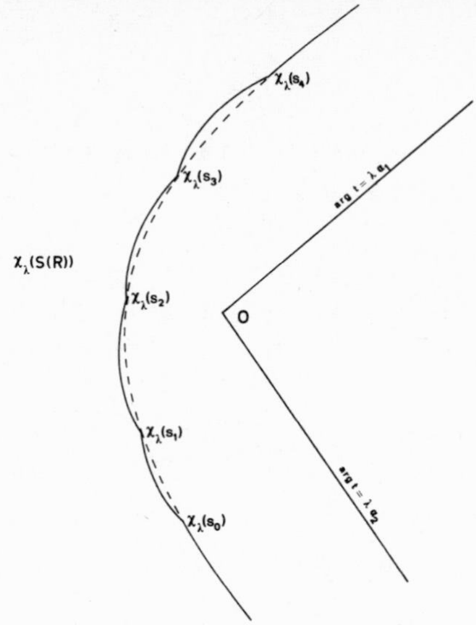

Fig. 2.
Sketch of $\chi_\lambda(S(R))$ for $\lambda > \kappa$, $N = 4$.
—: the curve $\chi_\lambda \circ \sigma$ --: the curve $\chi_\lambda \circ \sigma^*$

With (9.8) and (9.9) this may be rewritten in the form:

$$\alpha_\lambda^*(\xi_h) + \varepsilon_h^+ \leq \arg(t - t_h) \leq \alpha_\lambda^*(\xi_h) + \pi + \varepsilon_h^- \, . \tag{9.16}$$

Next, suppose that $\kappa \leq 1$ and $\lambda \in [\kappa, 1]$. In that case $\alpha_\lambda(\xi)$ is a monotonic nondecreasing function of arg $\sigma(\xi)$ on the intervals: $(-\infty, \xi_0)$, (ξ_N, ∞), and (ξ_h, ξ_{h+1}) $(h \in \{0,1,..,N-1\})$. Accordingly, we have

$$\alpha_\lambda^+ = \max \{\lambda\alpha_2 - \pi, \max_{h \in \{0,..,N-1\}} \alpha_\lambda^+(s_h)\}, \tag{9.17}$$

and

$$\alpha_\lambda^- = \min \{\lambda\alpha_1, \min_{h \in \{1,..,N\}} \alpha_\lambda^-(s_h)\}. \tag{9.18}$$

Using (9.7) - (9.11) we derive the inequalities

$$\lambda\alpha_2 - \pi \leq \alpha_\lambda^+ < \lambda\alpha_2 - \pi + (1 - \tfrac{\rho}{\kappa})\pi \tag{9.19}$$

and

$$\lambda\alpha_1 - (1 - \tfrac{\rho}{\kappa})\, \pi < \alpha_\lambda^- \leq \lambda\alpha_1 . \tag{9.20}$$

Let us assume that

$$\rho > \tfrac{1}{2}. \tag{9.21}$$

Since $\kappa > \frac{1}{2}$ this is not inconsistent with (9.5) . Then it follows that $\lambda < 2\rho$ for all $\lambda \in [\kappa, 1]$. Hence we derive that

$$\lambda\alpha_2 - \pi + (1 - \tfrac{\rho}{\kappa})\pi < \lambda\alpha_1 + \pi - (1 - \tfrac{\rho}{\kappa})\pi.$$

Combining this with (9.17) and (9.18) we conclude that

$$\alpha_\lambda^- \leq \alpha_\lambda^+ < \alpha_\lambda^- + \pi,$$

and consequently, (9.15) is not an empty statement.

Further, we wish to show that ρ can be chosen in such a way that, for all $k \in k_2(A)$, the following identity holds

$$[\tfrac{1}{k}\,\alpha_k^-, \tfrac{1}{k}\,\alpha_k^+] \cap \Sigma_k(A) = [\alpha_1, \alpha_2 - \tfrac{\pi}{k}] \cap \Sigma_k(A) \tag{9.22}$$

To this end we define, for all $k \in k(A)$,

$$\begin{aligned} \beta_k &= \max\{\beta \in \Sigma_k(A) : \beta < \alpha_1\} \\ \gamma_k &= \min\{\beta \in \Sigma_k(A) : \beta > \alpha_2 - \tfrac{\pi}{k}\} \end{aligned} \tag{9.23}$$

Now choose $\rho < \kappa$ so large that for all $k \in k_2(A)$ the inequality

$$(1 - \tfrac{\rho}{\kappa})\pi < k \min\{\alpha_1 - \beta_k, \gamma_k - (\alpha_2 - \tfrac{\pi}{k})\} \tag{9.24}$$

is satisfied. Then it is readily apparent from (9.19) and (9.20) that, for all $k \in k_2(A)$, we have

$$\beta_k < \tfrac{1}{k}\,\alpha_k^- \quad \text{and} \quad \tfrac{1}{k}\,\alpha_k^+ < \gamma_k,$$

and (9.22) follows.

So far, we have not verified whether $s \in S(R)$ implies $s - 1 \in S(R)$. If $\kappa > 1$ this is obviously the case, for then we have $\alpha_1^+ = \alpha_1 < \pi < \alpha_2 = \alpha_1^- + \pi$. If, on the other hand, $\kappa \leq 1$, the desired property is obtained by imposing on ρ the additional condition

$$(1 - \tfrac{\rho}{\kappa})\pi < \min\{\alpha_1, 2\pi - \alpha_2\}. \tag{9.25}$$

(If $\{0, \pi\} \subset \Sigma_1(A)$ then (9.25) is included in (9.24)). In both cases we now have

$$\pi \in (\alpha_1^+, \alpha_1^- + \pi). \tag{9.26}$$

Thus we have proved that, if ρ is any number satisfying the conditions (9.5) , (9.21) , (9.24) and (9.25) , then the regions $S(R) = S_\rho(R)$ defined above for all $R > 0$, form a set of S-proper regions.

Now suppose that $S' = S[\alpha_1', \alpha_2']$ is a left subsector of S. We use the same notation as before, adding a prime whenever a symbol is used in reference to S'. Thus, in the manner described above, we can define a set of S'-proper regions $\{S'(R), R > 0\}$. Let us take a positive number R and consider the region S'(R). To begin with, we shall assume that the values of α_1' and α_2' do not differ very much from those of α_1 and α_2, respectively. More precisely, we assume that, for all $k \in k_1(A)$, the following inequalities hold

$$\alpha_1' - \beta_k < \frac{\alpha_1 - \beta_k}{1 - k/\kappa}, \tag{9.27}$$

and

$$\gamma_k + \frac{\pi}{k} - \alpha_2' < \frac{\gamma_k + \frac{\pi}{k} - \alpha_2}{1 - k/\kappa}, \tag{9.28}$$

where β_k and γ_k are defined by (9.23) .

Let $k \in k_1(A)$, $\beta \in \Sigma_k(A)$ and suppose there is an integer $h \in \{1, .., N-1\}$ such that

$$\beta = \frac{1}{k} \alpha_k^*(\xi_h).$$

Let $R' \geq R$. The real number ξ_h determines a point s_h on the boundary of S(R'). By (9.3)

$$\arg s_h = \frac{\kappa\alpha_1 - k\beta}{\kappa - k}.$$

With (9.27) and (9.28) it follows that

$$\alpha_1' < \arg s_h < \alpha_2'.$$

(Note that $\gamma_k \leq \beta \leq \beta_k$ since $\alpha_k^*(\xi) \in (k\alpha_2 - \pi, k\alpha_1)$ for all $\xi \in (-\infty, \infty)$.)

Hence, by choosing a sufficiently large number R', we may achieve that

$$s_h \in S'(R).$$

Suppose that R' is so large that this is true for all $h \in \{1, .., N-1\}$ with the property that

$$\frac{1}{k} \alpha_k^*(\xi_h) \in \Sigma_k(A) \text{ for some } k \in k_1(A).$$

(Obviously, R' depends on R) We now define a closed region $\widetilde{S}(R)$ by

$$\widetilde{S}(R) = S'(R) \cup S(R').$$

It is easily seen that the regions $\tilde{S}(R)$ form another set of S-proper regions. They differ in many ways from the regions S(R) defined above. Thus, for instance, $\chi_\lambda(\tilde{S}(R))$ is usually not convex for any $\lambda > 0$. On the other hand, the analogue of (9.14) still holds in those cases that are of interest here. To see this, let us consider one particular region $\tilde{S}(R)$. For all $\lambda > 0$ we define real numbers $\tilde{\alpha}_\lambda^+$ and $\tilde{\alpha}_\lambda^-$ in analogy with α_λ^+ and α_λ^- above. We investigate the following three cases.

a. $\lambda \leq \rho$

Since $\kappa < \kappa'$ we may choose $\rho' \geq \rho$. Then, both $\chi_\lambda(S'(R))$ and $\chi_\lambda(S(R'))$ are convex. Using (9.12) and (9.13) we find that

$$\lambda\alpha_1 \leq \tilde{\alpha}_\lambda^+ \leq \lambda\alpha_1'$$

and

$$\lambda\alpha_2' - \pi \leq \tilde{\alpha}_\lambda^- \leq \lambda\alpha_2 - \pi.$$

Hence it follows that

$$0 < \tilde{\alpha}_\lambda^+ - \tilde{\alpha}_\lambda^- < \pi. \tag{9.29}$$

Moreover, it is easily seen that, for all $k \in k_1(A)$, the following relation holds

$$[\tfrac{1}{k}\tilde{\alpha}_k^-, \tfrac{1}{k}\tilde{\alpha}_k^+] \cap \Sigma_k(A) \subset [\alpha_2' - \tfrac{\pi}{k}, \alpha_1'] \cap \Sigma_k(A) \tag{9.30}$$

b. $\kappa \leq \lambda \leq \min(\rho', 1)$.

With the help of (9.12), (9.13), (9.19) and (9.20) we derive the inequalities

$$\min\{\lambda\alpha_1', \lambda\alpha_2 - \pi\} \leq \tilde{\alpha}_\lambda^+ \leq \max\{\lambda\alpha_1', \lambda\alpha_2 - \tfrac{\rho}{\kappa}\pi\}$$

and

$$\min\{\lambda\alpha_2' - \pi, \lambda\alpha_1 - (1 - \tfrac{\rho}{\kappa})\pi\} \leq \tilde{\alpha}_\lambda^- \leq \max\{\lambda\alpha_2' - \pi, \lambda\alpha_1\}.$$

By checking the different possibilities, one readily verifies that, if ρ is chosen in such a way that

$$\tfrac{\rho}{\kappa} \geq \tfrac{1}{\pi}\max\{\alpha_2 - \alpha_2', \alpha_1' - \alpha_1\},$$

then the inequality (9.29) holds again.

By the same method the following relation is found to be valid for all $k \in k_2(A)$ with the property that $k \leq \rho'$,

$$[\tfrac{1}{k}\tilde{\alpha}_k^-, \tfrac{1}{k}\tilde{\alpha}_k^+] \cap \Sigma_k(A) \subset ([\alpha_2' - \tfrac{\pi}{k}, \alpha_2 - \tfrac{\pi}{k}] \cup [\alpha_1, \alpha_1']) \cap \Sigma_k(A). \tag{9.31}$$

Here we have used (9.24) .

c. $\kappa' \leq \lambda \leq 1$.

In this case we have

$$\lambda\alpha_2' - \pi \leq \tilde{\alpha}_\lambda^+ \leq \max\{\lambda\alpha_2 - \frac{\rho}{\kappa}\pi, \lambda\alpha_2' - \frac{\rho'}{\kappa'}\pi\}$$

and

$$\min\{\lambda\alpha_1 - (1-\frac{\rho}{\kappa})\pi,\ \lambda\alpha_1' - (1-\frac{\rho'}{\kappa'})\pi\} \leq \tilde{\alpha}_\lambda^- \leq \lambda\alpha_1'.$$

Assuming that ρ and ρ' have been chosen in such a way that

$$\frac{\rho'}{\kappa'} \geq \frac{\rho}{\kappa} ,$$

we find that

$$\lambda\alpha_2' - \pi \leq \tilde{\alpha}_\lambda^+ \leq \lambda\alpha_2 - \frac{\rho}{\kappa}\pi$$

and

$$\lambda\alpha_1 - (1-\frac{\rho}{\kappa})\pi \leq \tilde{\alpha}_\lambda^- \leq \lambda\alpha_1' .$$

Hence we deduce that

$$0 \leq \tilde{\alpha}_\lambda^+ - \tilde{\alpha}_\lambda^- < \pi.$$

Utilizing (9.24) we find, for all $k \in k_2(A)$ such that $k \geq \kappa'$,

$$[\frac{1}{k}\tilde{\alpha}_k^-, \frac{1}{k}\tilde{\alpha}_k^+] \subset (\beta_k, \gamma_k)$$

It follows from the definitions of β_k and γ_k (cf. (9.23)) that

$$[\frac{1}{k}\tilde{\alpha}_k^-, \frac{1}{k}\tilde{\alpha}_k^+] \cap \Sigma_k(A) \subset [\alpha_1, \alpha_2 - \frac{\pi}{k}] \cap \Sigma_k(A). \tag{9.32}$$

The relations (9.30) - (9.32) have the following consequences. Suppose that $\alpha_1, \alpha_1', \alpha_2$ and α_2' have been chosen in such a way that, in addition to conditions (9.27) and (9.28), (4.7) is valid for a given $k \in k(A)$. Utilizing (9.12), (9.13) and (9.22) one readily verifies that, in all three cases, we have

$$[\frac{1}{k}\tilde{\alpha}_k^-, \frac{1}{k}\tilde{\alpha}_k^+] \cap \Sigma_k(A) \subset [\frac{1}{k}\alpha_k^-, \frac{1}{k}\alpha_k^+] \cap \Sigma_k(A) \tag{9.33}$$

Thus, for instance, if the latter intersection is empty, the same is true of the first.

Now, if conditions (9.27) and (9.28) are not both satisfied, then one can always choose two finite sequences $\{\alpha_1^{(0)}, .., \alpha_1^{(n)}\}$ and $\{\alpha_2^{(0)}, .., \alpha_2^{(n)}\}$, where $n \in \mathbb{N}$, $\alpha_1^{(0)} = \alpha_1'$, $\alpha_1^{(n)} = \alpha_1$, $\alpha_2^{(0)} = \alpha_2'$ and $\alpha_2^{(n)} = \alpha_2$, such that, for all $i \in \{0,..,n-1\}$ and all $k \in k_1(A)$, the following inequalities hold

$$\alpha_1^{(i+1)} - \beta_k \leq \alpha_1^{(i)} - \beta_k < \frac{\alpha_1^{(i+1)} - \beta_k}{1-k/\kappa'} ,$$

and

$$\gamma_k + \frac{\pi}{k} - \alpha_2^{(i+1)} \leq \gamma_k + \frac{\pi}{k} - \alpha_2^{(i)} < \frac{\gamma_k + \frac{\pi}{k} - \alpha_2^{(i+1)}}{1-k/\kappa'} .$$

Repeated application of the procedure described above then yields a sequence of positive numbers R_i and one of corresponding $S^{(i)}$-proper regions $S^{(i)}(R_i)$, where $S^{(i)} = S[\alpha_1^{(i)}, \alpha_2^{(i)}]$, $i = 0,1,..,n$, $R_0 = R$.
By defining

$$\widetilde{S}(R) = \bigcup_{i \in \{0,..,n\}} S^{(i)}(R_i)$$

one again obtains a suitable set of S-proper regions $\{\widetilde{S}(R), R > 0\}$. We shall not enter further into the details of this construction.

§*9.3. The case* $d = 0$.

Let S and A be defined as in §9.2 . Let $i \in \{1,..,\ell\}$ and suppose that the conditions of proposition 4.3 are satisfied. We assume that the canonical matrix A^c contains a 'block' B of the form

$$B = e^{q(s-1)-q(s)}(1 - \frac{1}{s})^G$$

where

$$\left.\begin{aligned} &q(s) = \sum_{ph=1}^{p} \mu_h s^h , \quad \mu_h \in ¢ , \\ &\text{and} \\ &G = \gamma I + N, \quad \gamma \in ¢, \quad N \text{ is a nilpotent } n \times n \text{ matrix.} \end{aligned}\right\} \tag{9.34}$$

Let Δ_B denote the left difference operator corresponding to B. The matrix function Y_0 defined by

$$Y_0(s) = e^{q(s)} s^G \tag{9.35}$$

is a fundamental matrix of the homogeneous linear difference equation

$$\Delta_B \, y(s) = 0. \tag{9.36}$$

As the constant μ_1 is determined modulo $2\pi i$ by B, (9.35) actually represents an infinite number of matrix solutions of (9.36). We shall choose a convenient one later on.

Let $g \in [k_{i-1}, k_i]$, $a \in \mathbb{R}$ and $r \in \mathbb{R}$. Let $R \geq 1$ and $f \in B_{g,a,r}(S(R))$, where $S(R)$ is an S-proper region of the form described in §9.2.

We define a linear mapping Λ on $B_{g,a,r}(S(R))$ by the formula

$$\Lambda f(s) = -A(s)^{-1} f(s) - \int_{C_1(s)} Y_0(s) Y_0(\zeta - 1)^{-1} [1 - e^{2\pi i (s-\zeta)}]^{-1} f(\zeta) d\zeta +$$

$$- \int_{C_2(s)} Y_0(s) Y_0(\zeta - 1)^{-1} [1 - e^{-2\pi i (s-\zeta)}]^{-1} f(\zeta) d\zeta - \int_{C(s)} Y_0(s) Y_0(\zeta - 1)^{-1} f(\zeta) d\zeta,$$

$$s \in S(R). \tag{9.37}$$

We take the paths $C_1(s)$ and $C_2(s)$ to be half-lines from $s - \frac{1}{2}$ to infinity, the respective directional angles $\hat{\alpha}_1$ and $\hat{\alpha}_2$ being chosen in such a way that

$$\hat{\alpha}_1 \in [\alpha_1^+, \pi) \ , \ \hat{\alpha}_2 \in (\pi, \alpha_1^- + \pi].$$

(For the definition of α_1^+ and α_1^- and further details concerning these quantities the reader is referred to §9.2, cf. in particular (9.26)).
By (9.15) this implies that both paths are contained in $S(R)$. $C(s)$ denotes a path in $S(R)$ going from a fixed point (which may be infinite) to $s - \frac{1}{2}$.

In the remainder of this section it will be shown that, by a suitable choice of the directional angles $\hat{\alpha}_1$ and $\hat{\alpha}_2$, of the path $C(s)$ and of the determination of the constant μ_1, one can achieve that the mapping Λ defined in (9.37) has the required properties.

We shall use the following notation. We put

$$s - \tfrac{1}{2} = s' \ ,$$

and

$$\|f\|_{g,a,r}^{S(R)} = \|f\| \ .$$

For all $\gamma \in (0,1]$ and all $\nu \in \mathbb{R}$, $F_{\gamma,a,\nu}$ denotes the function defined by

$$F_{\gamma,a,\nu}(x) = e^{ax^\gamma} x^\nu \ , \quad x \in \mathbb{R}^+. \tag{9.38}$$

From now on, the symbol K will be used to denote any constant which is independent of s and R and of the particular choice of f.

To begin with, we derive a few simple estimates that will be needed below. From (9.35) it is obvious that the matrix functions $Y_0(s)Y_0(s')^{-1}$ and $Y_0(s)Y_0(s-1)^{-1}$ are bounded on $S(1)$. Hence it follows that

$$|Y_o(s)Y_o(\zeta-1)^{-1}| < K|Y_o(s')Y_o(\zeta)^{-1}|, \quad s \in S(1). \tag{9.39}$$

Similarly, for all $\nu \in \mathbb{R}$ we have

$$F_{g,a,\nu}(|s'|) \leq K\, F_{g,a,\nu}(|s|), \quad s \in S(1). \tag{9.40}$$

Observing that arg s is bounded on S one readily verifies that

$$|(\frac{s'}{\zeta})^G| \leq |\frac{s'}{\zeta}|^{\mathrm{Re}\gamma}|P(\log|\frac{s'}{\zeta}|)|, \quad s \in S(1), \zeta \in S(1), \tag{9.41}$$

where P is a matrix polynomial.

We first discuss the integrals along the paths $C_j(s)$ $(j=1,2)$, to be denoted by $I_j(s)$. It easily follows from the definition of the paths that there exists a constant K, which in this case is independent of ζ as well, such that the inequality

$$|1-e^{\pm 2\pi i(s-\zeta)}|^{-1} \leq K|e^{\mp 2\pi i(s'-\zeta)}|, \quad s \in S(1), \zeta \in C_j(s),\ j = 1 \text{ or } 2, \tag{9.42}$$

is satisfied. Here and below we use the upper sign when $j=1$, the lower sign when $j=2$. Putting

$$q(s) - \mu_1 s = \psi(s), \tag{9.43}$$

and using (9.37), (9.40), (9.41) and (9.42) we find that

$$|I_j(s)| \leq K \int_{C_j(s)} |e^{\{(\mu_1 \mp 2\pi i)(s'-\zeta) + \psi(s') - \psi(\zeta)} P(\log|\frac{s'}{\zeta}|)f(\zeta)d\zeta|, \ j = 1,2). \tag{9.44}$$

We distinguish the following two cases.

1. Re $\mu_1 > 0$. Then we take

$$\hat{\alpha}_1 = \alpha_1^+, \quad \hat{\alpha}_2 = \alpha_1^- + \pi .$$

(If $\alpha_2 - \alpha_1 < \pi$ this definition implies that $\hat{\alpha}_1 = \alpha_1$ and $\hat{\alpha}_2 = \alpha_2$, cf. (9.12) and (9.13)). In view of (9.22) and the assumptions (ii) and (iii), the interval $[\alpha_1^-, \alpha_1^+]$ contains at most one 1-singular direction of A. Let the determinations of μ_1 and of arg μ_1 be chosen in such a way that $\frac{\pi}{2}$ − arg μ_1 is the one 1-singular direction, if there is any, contained in $[\alpha_1^-, \alpha_1^+]$, or else the largest 1-singular direction that is smaller than α_1^-. Consequently we have

$$\alpha_1^+ - \frac{\pi}{2} < \alpha_1^+ + \arg(\mu_1 - 2\pi i) < \frac{\pi}{2}$$

and

$$\frac{3}{2}\pi < \alpha_1^- + \pi + \arg(\mu_1 + 2\pi i) < \alpha_1^- + \frac{3}{2}\pi.$$

Since $[\alpha_1^-,\alpha_1^+] \subset (0,\pi)$ it follows that

$$\mathrm{Re}\{e^{i\hat{\alpha}_j}(\mu_1 \mp 2\pi i)\} > 0 \qquad (j = 1,2). \tag{9.45}$$

2. Re $\mu_1 \le 0$. In that case we take

$$\hat{\alpha}_1 = \max\{\alpha_1^+, \tfrac{\pi}{2}\}\ ,\ \hat{\alpha}_2 = \min\{\alpha_1^- + \pi, \tfrac{3}{2}\pi\}.$$

Furthermore, we choose the determinations of μ_1 and of arg μ_1 so that arg $\mu_1 \in [\frac{\pi}{2}, \frac{3}{2}\pi]$ and $|\mathrm{Im}\ \mu_1|$ is as small as possible. (If $\mu_1 = 0$ we take $\arg(\mu_1 + 2\pi i) = \frac{\pi}{2}$ and $\arg(\mu_1 - 2\pi i) = \frac{3}{2}\pi$). Then we have

$$\pi < \arg(\mu_1 - 2\pi i) \le \tfrac{3}{2}\pi$$

and

$$\tfrac{\pi}{2} \le \arg(\mu_1 + 2\pi i) < \pi.$$

Hence it follows that

$$\tfrac{3}{2}\pi < \hat{\alpha}_j + \arg(\mu_1 \mp 2\pi i) < \tfrac{5}{2}\pi \qquad (j = 1,2),$$

and thus (9.45) is again valid.
It is easily seen that the function Ψ defined in (9.43) satisfies the conditions of lemma 8.12.
Utilizing (9.40) and (9.45) and applying lemma 8.12 (cases (i) and (ii)) we obtain the following estimate

$$\|I_j\|^{S(R)}_{g,a,r} \le K\ \|f\| \qquad (j = 1,2)\ ,\ R \ge 1,$$

provided

$$g < 1 \quad \text{or} \quad g = 1 \quad \text{and} \quad |a| < \min_{j=1,2}\ \mathrm{Re}\{e^{i\hat{\alpha}_j}(\mu_1 \mp 2\pi i)\}. \tag{9.46}$$

If, instead of $S(R)$, the regions $\tilde{S}(R)$ defined at the end of §9.2 are considered, then, in the foregoing discussion the numbers α_1^+ and α_1^- should be replaced by $\tilde{\alpha}_1^+$ and $\tilde{\alpha}_1^-$, respectively. This leads to a result analogous to that obtained above, provided that (4.7) holds.

We want to make a final observation concerning the choice of the paths $C_j(s)$. The following relation is easily seen to hold for all $s \in S(1)$

$$C_j(s-1) = C_j(s) - 1 \qquad (j = 1,2). \tag{9.47}$$

Next we consider the integrals $I(s)$ along the paths $C(s)$,

$$I(s) = \int_{C(s)} Y_o(s)Y_o(\zeta-1)^{-1}f(\zeta)d\zeta, \quad s \in S(R). \tag{9.48}$$

By (9.39)

$$|I(s)| \leq K \int_{C(s)} |Y_o(s')Y_o(\zeta)^{-1}f(\zeta)d\zeta|. \tag{9.49}$$

Let $k = \text{degr}\ \underline{q}$ (cf. §2 for the definitions of $\underline{q}$ and degr). This implies that $k \in k(A)$.

We shall distinguish the following cases.

Case I: $k \neq 0$ and $(g,|a|) < (k,c)$.

Case II, which is subdivided into
II.1: $(g,a) < (k,-b) \leq 0$ and II.2: $g = k = 0$, $r < r_o$.

Case III, which in its turn is subdivided into
III.1: $(g,a) > (k,b') \geq 0$ and III.2: $g = k = 0$, $r > r_o$.

Here c,b,b' and r_o are real numbers, to be specified later. Note that case I relates to values of k such that $k \geq k_i$, whereas cases II and III relate to values for which $k \leq k_{i-1}$.
From now on we put

$$\mu_k = \mu.$$

I. $k \neq 0$, $(g,|a|) < (k,c)$.

By substituting

$$s' = t^{1/k} \quad \text{and} \quad \zeta = \tau^{1/k}$$

into the right-hand side of (9.49) and putting

$$q(s) - \mu s^k = \psi(t), \tag{9.50}$$

and

$$\chi_k(C(s)) = \ell(t),$$

where χ_k is the mapping defined by: $\chi_k(\zeta) = \zeta^k$ (cf. §9.2), we obtain, using (9.41),

$$|I(s)| \leq K \int_{\ell(t)} |e^{\mu(t-\tau)+\psi(t)-\psi(\tau)}P(\tfrac{1}{k}\log|\tfrac{t}{\tau}|)f(\tau^{1/k})\tau^{(1-k)/k}d\tau|. \tag{9.51}$$

a) Let us first suppose that $k \geq \kappa$ (i.e. $k \in k_2(A)$). By (4.5) and (9.22) $\theta \in [\alpha_k^-, \alpha_k^+]$ implies that $\theta + \arg \mu \neq \frac{\pi}{2}$ for all possible determinations of $\arg \mu$. Hence it follows that there is at least one $\beta \in [\alpha_k^+, \alpha_k^- + \pi]$ with the property that $\cos(\beta + \arg \mu) > 0$. Let $\alpha \in [\alpha_k^+, \alpha_k^- + \pi]$ be determined so that

$$\operatorname{Re}(e^{i\alpha}\mu) = \max_{\beta \in [\alpha_k^+, \alpha_k^- + \pi]} \operatorname{Re}(e^{i\beta}\mu).$$

The preceding argument shows that

$$\operatorname{Re}(e^{i\alpha}\mu) > 0. \tag{9.52}$$

We now take $\ell(t)$ to be the half-line from t to infinity, with directional angle α. It follows from (9.15) that $\ell(t) \subset \chi_k(S(R))$ for all $t \in \chi_k(S(R))$ and, consequently, that $C(s) \subset S(R)$ for all $s \in S(R)$. It is readily verified that the function Ψ defined in (9.50) satisfies the conditions of lemma 8.12.

Further, we observe that

$$|f(\tau^{1/k})\tau^{(1-k)/k}| \leq \|f\| F_{g/k,a,(r+1-k)/k}(|\tau|), \qquad \tau \in \chi_k(S(R)). \tag{9.53}$$

Inserting this into (9.51), utilizing (9.40) and (9.52) and applying lemma 8.12 (cases (i) and (ii)) we obtain the estimate

$$\|I\|^{S(R)}_{g,a,r+1-k} \leq K \|f\|, \qquad R \geq 1,$$

provided

$$g < k, \text{ or } g = k \text{ and } |a| < c,$$

where

$$c = \max_{\beta \in [\alpha_k^+, \alpha_k^- + \pi]} \operatorname{Re}(e^{i\beta}\mu).$$

A similar result is found with regard to the regions $\tilde{S}(R)$, if the numbers α_k^+ and α_k^- are replaced by $\tilde{\alpha}_k^+$ and $\tilde{\alpha}_k^-$, respectively, provided that (4.7) holds. (In that case (9.33) is used).

In view of the exponential decrease of the integrand in (9.48), the following identity is an immediate consequence of Cauchy's theorem,

$$\int_{C(s-1)} Y_o(s)Y_o(\zeta-1)^{-1}f(\zeta)d\zeta = \int_{C(s)-1} Y_o(s)Y_o(\zeta-1)^{-1}f(\zeta)d\zeta.$$

Combining this with (9.47) one readily verifies that in this case the mapping Λ defined in (9.37) is both a left and a right inverse of Δ_B.
Moreover, if $S' = S[\alpha_1', \alpha_2']$ is a left subsector of S such that (4.7) is satisfied, and S'(R) is an S'-proper region of the same type as S(R), then it is easily seen from the definition of Λ that (4.8) is true.
b) Now suppose that $k < \kappa$ (i.e. $k \in k_1(A)$). The following two situations may arise.
(i) There is an $\alpha \in [\alpha_k^+, \alpha_k^- + \pi]$ such that $\mathrm{Re}(e^{i\alpha}\mu) > 0$. In this case the paths C(s) are defined in the same manner as previously and analogous results are thus obtained. In particular, the mapping Λ is again an inverse of the difference operator corresponding to B.
(ii) $\mathrm{Re}(e^{i\alpha}\mu) < 0$ for all $\alpha \in [\alpha_k^+, \alpha_k^- + \pi]$. This implies that the interval (α_k^-, α_k^+) or, equivalently, $(k\alpha_2 - \pi, k\alpha_1)$, contains a number β with the property that

$$\beta = \frac{\pi}{2} - \arg \mu \tag{9.54}$$

for a suitable determination of $\arg \mu$. Since $\alpha_k^*(\xi)$ is a continuous function of ξ, monotonically increasing from $k\alpha_2 - \pi$ to $k\alpha_1$ (cf. (9.3)), there must be a real number ξ such that $\beta = \alpha_k^*(\xi)$. Clearly $\frac{1}{k}\beta \in \Sigma_k(A)$. Thus, by (9.4) there is an integer $h \in \{1,..,N-1\}$ such that

$$\beta = \alpha_k^*(\xi_h). \tag{9.55}$$

Let

$$\chi_k(s_h) = t_h$$

and let $\ell(t)$ be defined as the directed segment from t_h to t. Since $\chi_k(S(R))$ is a convex set we know that $\ell(t) \subset \chi_k(S(R))$ for all $t \in \chi_k(S(R))$ and hence $C(s) \subset S(R)$ for all $s \in S(R)$. Combining (9.55) and (9.16) we find

$$\varepsilon_h^+ \leq \arg(t - t_h) - \beta \leq \pi + \varepsilon_h^-, \qquad t \in \chi_k(S(R)).$$

With (9.54), (9.10) and (9.11) it follows that

$$\cos\{\arg(t_h - t) + \mu\} \geq \min\{-\sin \varepsilon_h^-, \sin \varepsilon_h^+\} > 0. \tag{9.56}$$

Utilizing (9.51) ,(9.40), (9.53) and (9.56) and applying lemma 8.12 (cases (i) and (ii), with $\alpha = \arg(t_h - t)$) we obtain a result of the same form as in a) with the only difference that

$$c = \min_{t \in \chi_k(S(R))} \mathrm{Re}\{e^{i \arg(t_h - t)}\mu\} = |\mu| \min\{-\sin \varepsilon_h^-, \sin \varepsilon_h^+\}. \tag{9.57}$$

However, in contrast to the preceding cases, the mapping Λ is not a left inverse of Δ_B. By means of residue calculus it is readily verified that

$$\Lambda\Delta_B\, f(s) = f(s) + Y_0(s) \int_{s_h}^{s_h-1} Y_0(\zeta)^{-1} f(\zeta)d\zeta, \qquad s \in S(R). \tag{9.58}$$

Finally, suppose that $S' = S[\alpha_1', \alpha_2']$ is a left subsector of S such that (4.7) is satisfied and let $\{S'(R),\ R>0\}$ be a set of S'-proper regions of the same type as the regions S(R). We wish to prove that, in the case considered here, the statement made on p. 17 , immediately after proposition 4.3 , is true. This requires a slight modification of the S-proper regions and the mappings Λ. Instead of S(R) we now consider the regions $\tilde{S}(R)$. In order to simplify the discussion, we shall assume that (9.27) and (9.28) hold and, accordingly, $\tilde{S}(R)$ has the form

$$\tilde{S}(R) = S'(R) \cup S(R'), \qquad R > 0,$$

where R' is a positive number, depending on R. With respect to S(R') we shall use the same notation we introduced for S(R) above, whereas symbols referring to S'(R) will be supplied with a prime. Thus, in the manner described above, we can define linear mappings Λ and Λ' on $B_{g,a,r}(S(R'))$ and $B_{g,a,r}(S'(R))$, respectively. The difference between Λ and Λ' lies in the definition of the paths C(s) and C'(s). The first is defined as the directed segment from s_h to $s'(= s - \frac{1}{2})$, whereas C'(s) is the directed segment from the point s_h' on the boundary of S'(R) to s'.
Now let $f \in B_{g,a,r}(\tilde{S}(R))$. The restrictions of f to S'(R) and S(R'), will also be denoted by f. By Cauchy's theorem we have, for all $s \in S'(R) \cap S(R')$,

$$\Lambda' f(s) - \Lambda f(s) = Y_0(s) \int_{C'(s_h+\frac{1}{2})} Y_0(\zeta)^{-1} f(\zeta)d\zeta,$$

or, equivalently,

$$\Lambda' f(s) - \Lambda f(s) = Y_0(s) Y_0(s_h+\tfrac{1}{2})^{-1}\ I'(s_h+\tfrac{1}{2}),$$

where I' is defined analogously to I (cf. (9.48)).
We define a linear mapping $\tilde{\Lambda}$ on $B_{g,a,r}(\tilde{S}(R))$ as follows

$$\tilde{\Lambda} f(s) = \Lambda' f(s) \qquad \text{if } s \in S'(R),$$

$$\tilde{\Lambda} f(s) = \Lambda f(s) + Y_0(s) Y_0(s_h+\tfrac{1}{2})^{-1}\ I'(s_h+\tfrac{1}{2}) \qquad \text{if } s \in S(R').$$

Clearly, $\tilde{\Lambda}f$ is the analytic continuation into $\tilde{S}(R)$ of $\Lambda'f$, i.e. $\tilde{\Lambda}$ has the property expressed in (4.8). Hence it follows that $\tilde{\Lambda}$ is a right inverse of Δ_B. Utilizing (9.56) and the results obtained by the above described method with respect to Λ and Λ', one readily verifies that

$$\|\tilde{\Lambda}f\|^{\tilde{S}(R)}_{g,a,r+1-k} \leq K\ \|f\|^{\tilde{S}(R)}_{g,a,r}\ , \qquad R \geq 1,$$

provided that $(g,|a|) < (k,\tilde{c})$, where $\tilde{c} = |\mu|\ \min\{\varepsilon_h^-,(\varepsilon_h^-)',\ \varepsilon_h^+,(\varepsilon_h^+)'\}$.

II.1 $(g,a) < (k,-b) \leq 0$.

By a corollary to a theorem of Phragmén-Lindelöf (cf.theorem 3.3) $B_{g,a,r}(S(R))$ is reduced to $\{0\}$ when $g \geq \kappa$. Therefore it is sufficient to consider the case that $g < \kappa$.

Let

$$\mu_g = 0 \quad \text{if} \quad g > k, \quad \mu_g = \mu \quad \text{if} \quad g = k. \tag{9.59}$$

Substituting

$$s' = t^{1/g} \quad \text{and} \quad \zeta = \tau^{1/g}$$

into the right-hand side of (9.49), utilizing (9.41) and putting

$$q(s) - \mu_g s = \Psi(t),$$

and

$$\chi_g(C(s)) = \ell(t),$$

we find

$$|I(s)| \leq K \left|\int_{\ell(t)} e^{\mu_g(t-\tau)+\Psi(t)-\Psi(\tau)} P\left(\frac{1}{g}\log\left|\frac{t}{\tau}\right|\right) f(\tau^{1/g})\tau^{(1-g)/g} d\tau\right|. \tag{9.60}$$

As path of integration $\ell(t)$ we take the half-line from t to infinity with directional angle arg t. (9.6) implies that $\ell(t) \subset \chi_g(S(R))$ and thus $C(s) \subset S(R)$. It is not difficult to verify that the resulting function $I(s)$ is holomorphic in int $S(R)$.

Observe that

$$|f(\tau^{1/g})\tau^{(1-g)/g}| \leq \|f\| F_{1,a,(r+1-g)/g}(|\tau|), \qquad \tau \in \chi_g(S(R)). \tag{9.61}$$

Inserting this into (9.60), utilizing (9.40) and applying lemma 8.12 (cases

(ii) and (iii), with $\beta = 0$), we obtain the following estimate

$$\|I\|^{S(R)}_{g,a,r+1-g} \leq K\ \|f\|, \qquad R \geq 1,$$

provided that

$$g \geq k \quad \text{and} \quad a < \min_{\beta \in [g\alpha_1, g\alpha_2]} \operatorname{Re}(e^{i\beta}\mu_g),$$

i.e. $b = -\min_{\beta \in [k\alpha_1, k\alpha_2]} \operatorname{Re}(e^{i\beta}\mu) = -\min_{\beta \in [\alpha_k^+, \alpha_k^- + \pi]} \operatorname{Re}(e^{i\beta}\mu)$.

Note that this result is slightly stronger than what was announced above. Indeed, if $\min_{\beta \in [\alpha_k^+, \alpha_k^- + \pi]} \operatorname{Re}(e^{i\beta}\mu) > 0$, then also nonnegative values of a are allowed. In that case the function I(s) defined here coincides with that defined in I.b(i), provided

$$|a| < \max_{\beta \in [\alpha_k^+, \alpha_k^- + \pi]} \operatorname{Re}(e^{i\beta}\mu).$$

The linear mapping Λ has the same properties in this case as in cases I.a and I.b(i). In particular, it is an inverse of Δ_B.

II.2 $g = k = 0$, $r < -r_o$.

In this case we have

$$Y_o(s) = s^G, \qquad s \in S(1).$$

Hence, by (9.41)

$$|Y_o(s')Y_o(\zeta)^{-1}| \leq |\tfrac{s'}{\zeta}|^{\operatorname{Re}\gamma} |P(\log|\tfrac{s'}{\zeta}|)|, \qquad s \in S(1),\ \zeta \in S(1).$$

Inserting this into (9.48) we obtain

$$|I(s)| \leq K \int_{C(s)} |\tfrac{s'}{\zeta}|^{\operatorname{Re}\gamma} |P(\log|\tfrac{s'}{\zeta}|) f(\zeta) d\zeta|, \qquad s \in S(R),\ R \geq 1.$$

As path of integration C(s) we take the half-line from s' to infinity with directional angle args.

Observing that

$$|f(\zeta)| \leq \|f\|\ |\zeta|^r$$

and putting

$$\zeta = x\,s', \qquad x \in (1,\infty),$$

we find

$$|I(s)| \le K\,\|f\|\,|s'|^{r+1} \int_1^\infty x^{r-\mathrm{Re}\gamma}|P(-\log x)|dx.$$

If follows immediately that

$$\|I\|^{S(R)}_{o,o,r+1} \le K\,\|f\|, \qquad R \ge 1,$$

provided

$$r < -r_o,$$

where $r_o = -\operatorname{Re}\gamma + 1$.

The linear mapping Λ has the same properties in this case as in the foregoing one.

III.1. $(g,a) > (k,b') \ge 0$.

a) Let us first suppose that $g \le \rho$, where ρ is the positive number introduced in § 9.2.

We use the same notation as in II.1. Obviously, (9.60) and (9.61) are again valid. It is apparent from the considerations held in §9.2 that the sequence $\{\xi_0,\xi_1,..,\xi_N\}$ may contain arbitrary real numbers in addition to those referred to in (9.4). Therefore, we may assume that there is an integer $h \in \{0,..,N\}$ such that

$$\xi_h = 0.$$

The corresponding point s_h on the curve σ^* is the only point in $S(R)$ at a distance R from the origin. It follows from (9.2) that

$$\arg s_h = \alpha_1 + \frac{\pi}{2\kappa}\,. \tag{9.62}$$

Let

$$\chi_g(s_h) = t_h$$

and let $\ell(t)$ be the directed segment from t_h to t. Since $g \le \rho$, $\chi_\rho(S(R))$ is convex, which implies that $\ell(t) \subset \chi_g(S(R))$ and, consequently, $C(s) \subset S(R)$. By (9.16)

$$\varepsilon_h^+ \le \arg(t-t_h) - \alpha_g^*(0) \le \pi + \varepsilon_h^-\,, \qquad t \in \chi_g(S(R)). \tag{9.63}$$

From (9.3) and (9.62) we deduce that

$$\alpha_g^*(0) = \arg t_h - \frac{\pi}{2} .$$

Combining this with (9.63) we find

$$-\frac{\pi}{2} + \varepsilon_h^+ \leq \arg(t - t_h) - \arg t_h \leq \frac{\pi}{2} + \varepsilon_h^- , \qquad t \in \chi_g(S(R)).$$

In view of (9.10) and (9.11) this implies that

$$|\arg(t - t_h) - \arg t_h| \leq \frac{\pi}{2} - \min\{\varepsilon_h^+, -\varepsilon_h^-\} < \frac{\pi}{2} . \tag{9.64}$$

Utilizing (9.60) , (9.40),(9.61) and (9.64) and applying lemma 8.12 (case (iv), with $\alpha = \arg(t_h - t)$) we obtain the estimate

$$\|I\|_{g,a,r+1-g}^{S(R)} \leq K \|f\|, \qquad R \geq 1,$$

provided that

$$g > k \quad \text{and} \quad a > 0, \quad \text{or} \quad g = k \quad \text{and} \quad a > b',$$

where

$$b' = [\sin(\min\{\varepsilon_h^+, -\varepsilon_h^-\})]^{-1} \max\{0, -\min_{t \in \chi_k(S(R))} \mathrm{Re}\{e^{i \arg(t_h - t)}\mu\}\}.$$

b) Now suppose that $g > \rho$.

Using (9.35) and putting

$$s' = t^{1/\rho} , \quad \zeta = \tau^{1/\rho} ,$$

and

$$\ell(t) = \chi_\rho(C(s)),$$

we may rewrite (9.49) in the form

$$|I(s)| \leq K \int_{\ell(t)} |e^{q(t^{1/\rho}) - q(\tau^{1/\rho})} (\frac{t}{\tau})^{\frac{1}{\rho}G} f(\tau^{1/\rho}) \tau^{1/\rho - 1} d\tau|, \qquad s \in S(R). \tag{9.65}$$

Let s_h be defined in the same manner as in a) and let

$$t_h = \chi_\rho(s_h).$$

We choose for $\ell(t)$ the directed segment from t_h to t. Since $\chi_\rho(S(R))$ is convex, $\ell(t) \subset \chi_\rho(S(R))$ and hence $C(s) \subset S(R)$. We shall consider the integration variable τ as a function of the parameter x, in the following manner,

$$\tau(x) = t_h - e^{i \arg(t - t_h)} x, \qquad x \in (x_0, 0),$$

where $x_0 = -|t - t_h|$.

Further, we define real functions u, φ_1, φ_2 and φ on $(x_0,0)$ as follows,

$$u(x) = |\tau(x)|,$$

$$\varphi_1(x) = \mathrm{Re}\{q(\tau(x)^{1/\rho})\} - a\, u(x)^{g/\rho},$$

$$\varphi_2(x) = \frac{1}{\rho}(\mathrm{Re}\,\gamma - r - 1 + \rho)\log u(x), \tag{9.66}$$

and

$$\varphi(x) = \varphi_1(x) + \varphi_2(x).$$

Observing that

$$|f(\tau(x)^{1/\rho})| \leq \|f\| \exp\{a\, u(x)^{g/\rho}\}\, u(x)^{r/\rho},$$

utilizing (9.65),(9.40),(9.41) and the definitions given above, we obtain an inequality of the form

$$|I(s)|\{F_{g,a,r+1-\rho}(|s|)\}^{-1} \leq K\|f\| \int_{x_0}^{o} |e^{\varphi(x_0)-\varphi(x)} P[\frac{1}{\rho}\{\log \frac{u(x_0)}{u(x)}\}]dx|. \tag{9.67}$$

Since $\ell(t) \subset \chi_\rho(S(R))$ we have

$$u(x) \geq R^\rho, \qquad x \in (x_0,0). \tag{9.68}$$

Analogously to (8.13) we find that

$$u'(x) = -\cos\{\arg(t - t_h) - \arg \tau(x)\}, \qquad x \in (x_0,0)$$

It is easily seen that (9.64) is again valid.
Consequently,

$$|\arg(t - t_h) - \arg \tau(x)| \leq |\arg(t - t_h) - \arg t_h| \leq \frac{\pi}{2} - \min\{\varepsilon_h^+, -\varepsilon_h^-\},$$

$$x \in (x_0,0).$$

Hence it follows that

$$u'(x) \leq -\sin(\min\{\varepsilon_h^+, -\varepsilon_h^-\}), \qquad x \in (x_0,0). \tag{9.69}$$

From the definition of q in (9.34) we deduce

$$\frac{d}{dx} q(\tau(x)^{1/\rho}) = -\sum_{pj=1}^{p} \frac{j}{\rho} \mu_h e^{i\arg(t-t_h)} \tau(x)^{j/\rho-1}, \qquad x \in (x_0,0).$$

This implies

$$\frac{d}{dx} \mathrm{Re}\; q(\tau(x)^{1/\rho}) \geq \frac{g}{\rho} u(x)^{g/\rho-1}(-|\mu_g| - c\, R^{-\varepsilon}), \qquad x \in (x_0,0),$$

where μ_g is defined by (9.59), c is a positive number independent of x and t, $\varepsilon = \frac{1}{p}$ if $g = k$ and $\varepsilon = g - k$ if $g > k$. Using (9.69) we conclude that

$$\varphi_1'(x) \geq \frac{g}{\rho} u(x)^{g/\rho-1}[-|\mu_g| + a \sin(\min\{\varepsilon_h^+, -\varepsilon_h^-\}) - c\, R^{-\varepsilon}], \qquad x \in (x_0,0). \tag{9.70}$$

Now suppose that

$$a > |\mu_g| [\sin(\min\{\varepsilon_h^+, -\varepsilon_h^-\})]^{-1}.$$

Let $R_o \geq 1$ be chosen so large that

$$-|\mu_g| + a \sin(\min\{\varepsilon_h^+, -\varepsilon_h^-\}) - c\, R_o^{-\varepsilon} > 0$$

and let $R \geq R_o$. Then it follows from (9.70) and (9.68) that there exists a positive number δ such that

$$\varphi_1'(x) \geq \delta, \qquad x \in (x_o, 0).$$

Furthermore, it is apparent from (9.66) that the functions φ_2 and φ_2' satisfy inequalities of the form (8.5) and (8.6), respectively.

Applying lemma 8.1 to the integral on the right-hand side of (9.67) we obtain the estimate

$$\|I\|^{S(R)}_{g,a,r+1-\rho} \leq K\|f\|, \qquad R \geq R_o,$$

provided that

$$g > k \text{ and } a > 0 \quad \text{or} \quad g = k \text{ and } a > b',$$

where

$$b' = [\sin(\min\{\varepsilon_h^+, -\varepsilon_h^-\})]^{-1} |\mu|.$$

Obviously, R_o is independent of r.

III.2 $g = k = 0$, $r > r_o$.

Using the same notation as in the previous case we now have

$$|I(s)| \leq K \int_{\ell(t)} |\tfrac{t}{\tau}|^{\mathrm{Re}\gamma/\rho} |P(\tfrac{1}{\rho} \log |\tfrac{t}{\tau}|) f(\tau^{1/\rho}) \tau^{1/\rho - 1}\, d\tau| \tag{9.71}$$

Let $\ell(t)$ again be the directed segment from t_h to t, where t_h is defined as in III.1.b).

We put

$$\tau = t(1 + e^{i\alpha(t)} x), \qquad x \in (0, x(t)), \tag{9.72}$$

where $\alpha(t) = \arg(t_h - t) - \arg t$, $x(t) = |1 - \frac{t_h}{t}|$. (The direction in which the path is described is reversed by this definition but that is immaterial here).

Further, we have

$$|f(\tau^{1/\rho})\ \tau^{1/\rho - 1}| \leq \|f\|\ |\tau|^{(r+1-\rho)/\rho}, \qquad \tau \in \chi_\rho(S(R)). \tag{9.73}$$

Inserting (9.72) and (9.73) into (9.71) and putting

$$\mathrm{Re}\gamma - 1 = r_o$$

we find

$$|I(s)| \leq K\,\|f\|\,|t|^{(r+1)/\rho} \int_0^{x(t)} |1+e^{i\alpha(t)}x|^{(r-r_o-\rho)/\rho}|P(\tfrac{1}{\rho}\log|1+e^{i\alpha(t)}x|)|\,dx.$$

As we pointed out before, (9.64) is again valid. Consequently, $|t-t_h| < |t|$ and thus

$$x(t) < 1, \qquad t \in \chi_\rho(S(R)).$$

Clearly, the following inequalities hold for all $x \in (0,1)$ and all $t \in \chi_\rho(S(R))$,

$$|1+e^{i\alpha(t)}x|^{(r-r_o-\rho)/\rho} \leq (1+x)^{(r-r_o-\rho)/\rho} \qquad \text{if } r > r_o + \rho,$$

$$|1+e^{i\alpha(t)}x|^{(r-r_o-\rho)/\rho} \leq (1-x)^{(r-r_o-\rho)/\rho} \qquad \text{if } r < r_o + \rho,$$

and

$$|\log|1+e^{i\alpha(t)}x|\,| \leq -\log(1-x).$$

If follows immediately that

$$\|I\|_{r+1}^{S(R)} \leq K\,\|f\|, \qquad R \geq 1,$$

provided $r > r_o$.

The linear mapping Λ we thus obtain in each of the three subcases of case III is a right, but not a left inverse of Δ_B. One readily verifies that (9.58) holds again, if s_h is the point of σ^* determined by (9.62).

Finally, suppose that $S' = S[\alpha_1', \alpha_2']$ is a left subsector of S such that (4.7) holds. Let $\{\widetilde{S}(R), R > 0\}$ be a set of S-proper regions of the type described at the end of §9.2. Under the conditions that we have imposed on g, a and r above, we can define a linear mapping $\widetilde{\Lambda}$ on $B_{g,a,r}(\widetilde{S}(R))$ analogously to that defined in case I.b(ii) and having the same properties, in particular the one expressed by (4.8).

§9.4. *The case* $d < 0$.

We shall deal with a slightly more general situation than is required for the proof of proposition 4.3.

Let $S = S[\alpha_1, \alpha_2]$, where $\alpha_1 \in (0,\pi), \alpha_2 \in [\pi, 2\pi)$, and let $\{S(R), R > 0\}$ be an arbitrary set of S-proper regions. Suppose that B is an $n \times n$ matrix function of the form

$$B(s) = s^{-d}\,\tilde{B}(s), \qquad s \in S(1),$$

where d is a negative real number, and $\tilde{B}$ is a matrix function with the property that both $\tilde{B}$ and $\tilde{B}^{-1}$ are continuous on S(1) and holomorphic in its interior. Δ_B will denote the difference operator corresponding to B.

Let $g \in [0,1]$, $a \in \mathbb{R}$ and $r \in \mathbb{R}$. Let $R \geq 1$. We define a linear mapping Λ on $B_{g,a,r}(S(R))$ as follows,

$$\Lambda f(s) = -\sum_{h=o}^{\infty} B(s)^{-1}B(s-1)^{-1}..B(s-h)^{-1}f(s-h), \qquad f \in B_{g,a,r}(S(R)). \tag{9.74}$$

(This definition is equivalent to (7.2) if we take for Y_o a fundamental matrix of the equation $\Delta_B\, y = 0$).
There exists a positive number b such that

$$\sup_{s \in S(R_o)} |\tilde{B}(s)^{-1}| = b.$$

Thus

$$|\Lambda f(s)| \leq \sum_{h=o}^{\infty} b^{h+1}(|s|\,|s-1|..|s-h|)^d |f(s-h)|. \tag{9.75}$$

We have

$$|f(s-h)| \leq \|f\| \exp(a|s-h|^g)\; |s-h|^r, \qquad s \in S(1),\; h \in \mathbb{N} \cup \{0\}. \tag{9.76}$$

By the mean value theorem, there exists a number $\eta \in (0,h)$ such that

$$|s-h|^g - |s|^g = g|s-\eta|^{g-1}\cos\{\arg(s-\eta)\}h.$$

Since $|s-\eta| \geq 1$ if $s \in S(1)$, it follows that

$$\exp\{a(|s-h|^g - |s|^g)\} \leq K \exp(|a|h), \qquad s \in S(1),\; h \in \mathbb{N}. \tag{9.77}$$

Here and below, the capital K denotes any constant which is independent of s, h and R and of the particular choice of the function f.
Let $\alpha = \min\{\alpha_1, 2\pi - \alpha_2\}$. Then the following inequality holds for all $s \in S$ and all $h \in \mathbb{N}$,

$$|s-h|^2 \geq |s|^2 - 2|s|h\cos\alpha + h^2.$$

Hence we deduce that

$$|s-h| \geq |s| \sin\alpha, \qquad s \in S,\; h \in \mathbb{N}, \tag{9.78}$$

and, similarly,

$$|s-h| \geq h \sin \alpha, \qquad s \in S,\ h \in \mathbb{N}. \tag{9.79}$$

If $r<0$, it follows from (9.78) that

$$\left|\frac{s-h}{s}\right|^r \leq (\sin \alpha)^r, \qquad s \in S,\ h \in \mathbb{N}.$$

If on the other hand $r \geq 0$, and $s \in S(1)$, we find

$$\left|\frac{s-h}{s}\right|^r \leq (h+1)^r, \qquad s \in S(1),\ h \in \mathbb{N}.$$

Combining the last two inequalities we see that, in both cases,

$$\left|\frac{s-h}{s}\right|^r \leq K\, h^{|r|}, \qquad s \in S(1),\ h \in \mathbb{N}. \tag{9.80}$$

Furthermore, we conclude from (9.79) that

$$|s-1| \cdot \ldots \cdot |s-h| \geq h!\ (\sin \alpha)^h, \qquad s \in S(1),\ h \in \mathbb{N} \tag{9.81}$$

Utilizing (9.75) , (9.76) ,(9.77) , (9.80) and (9.81) we obtain, for all $s \in S(1)$, the inequality

$$e^{-a|s|^g} |s|^{-r-d} |\Lambda f(s)| \leq \|f\| \,[b + K \sum_{h=1}^{\infty} (h!)^d\, h^{|r|} \{b\, e^{|a|} (\sin \alpha)^d\}^h],$$

and thus

$$\|\Lambda f\|_{g,a,r+d}^{S(R)} \leq K \|f\|, \qquad R \geq 1.$$

From the definition in (9.74) and the estimates found above it follows immediately that Λ is an inverse of Δ_B. Observe that the definition of Λ is independent of the values of α_1 and α_2 and the shape of the regions $S(R)$. Hence it is clear that Λ has the property stated in (4.8) .

§9.5. Conclusions.

Returning to the original matrix A and the corresponding difference operator Δ, we come to the conclusion that the methods described in this section enable us to construct linear mappings $\Lambda_{g,a,r}^{S(R)}$ possessing the properties mentioned in proposition 4.3 and the additional property stated in (4.8) .

From the results obtained in §9.3 and §9.4 we can collect the following information concerning the constants c_i and b_i. If $i \neq \ell$, we may take for c_i any positive number smaller than the minimum of all constants c (cf. §9.3, case I) corresponding to blocks B of A with the property that $k(B) = \{k_i\}$.

In the case that $i = \ell$, and hence $k_i = 1$, we must take into account the additional condition (9.46).
Further, if $i \in \{2,..,\ell\}$ we may choose for b_i any number larger than the maximum of all constants b and b' (cf. §9.3, cases II and III) corresponding to blocks B with the property that $k(B) = \{k_{i-1}\}$. (The constant b_1 is of no importance since $k_o = 0$).
Clearly, the conditions imposed on a and r could be refined by distinguishing between positive and negative values of a and r. However, we do not need such precise results for our purpose.

§10. Proof of proposition 4.9.

We consider the case that $j = 1$ and $\alpha_1 = 0$. The other cases can be treated analogously.

Let $\beta \in (0, \frac{\pi}{2})$ be chosen in such a way that, for all $k \in k(A)$ such that $k \geq k_i$, and for all $\mu \in \mu_k(A)$,

$$\mathrm{Re}(\mu e^{i(k-1)\beta}) < 0 \tag{10.1}$$

and let $\alpha \in (0,\beta)$. The third assumption of proposition 4.9 implies that, for all $k \in k(A)$ such that $k \geq k_i$,

$$\Sigma_k(A) \subset \bigcup_{m \in \mathbb{Z}} \left(\frac{(2m-1)\pi}{k}, \frac{2m\pi}{k}\right).$$

Hence it follows that, with respect to the sector $S[\alpha,\alpha_2]$, the conditions of proposition 4.3 are satisfied. Let $\{S(R), R > 0\}$ be a set of $S[\alpha,\alpha_2]$-proper regions as defined in §9.2. By s_R we shall denote the point on the boundary of $S(R)$ with the property that $\arg s_R = \beta$. For all $R > 0$ we define a closed region $\tilde{S}(R)$ by

$$\tilde{S}(R) = S(R) \cup \{s \in \mathbb{C} : \mathrm{Im}\, s \geq \mathrm{Im}\, s_R\}.$$

It is easily seen that the set $\{\tilde{S}(R), R > 0\}$ is a set of S-proper regions.

Again we may assume that A is in canonical form. We consider one block B of the form (9.1) and denote the corresponding difference operator by Δ_B.

Let $r \in \mathbb{R}$, $g \in [k_{i-1}, k_i]$ and let a be a real number such that (4.6) is satisfied. There exist a real number v, a positive number R_o and linear mappings

$$\Lambda^{S(R)}_{g,a,r} : B_{g,a,r}(S(R)) \longrightarrow B_{g,a,r+v}(S(R)),$$

defined for all $R \geq R_o$ and possessing the properties mentioned in proposition 4.3, in particular $\Lambda^{S(R)}_{g,a,r}$ is a right inverse of Δ_B.

Let $f \in B_{g,a,r}(\tilde{S}(R))$. We define a linear mapping $\tilde{\Lambda}$ on $B_{g,a,r}(\tilde{S}(R))$ by means of the following relations

$$\tilde{\Lambda}f(s) = \Lambda^{S(R)}_{g,a,r} f(s) \qquad \text{if } s \in S(R)$$

and

$$\tilde{\Lambda}f(s) = B(s)^{-1}\{\tilde{\Lambda}f(s-1) - f(s)\}, \qquad s \in \tilde{S}(R). \tag{10.2}$$

Consider the regions $S_N(R)$ defined by

$$S_N(R) = \{s \in \tilde{S}(R) : \text{Re } s \leq N\}, \qquad N \in \mathbb{N},\ R > 0.$$

Let $N_o \in \mathbb{N}$ and suppose there exist positive numbers $\tilde{R}_o$ and $\tilde{K}$ such that, for all $N < N_o$ and all $R \geq \tilde{R}_o$,

$$||\tilde{\Lambda}f||^{S_N(R)}_{g,a,r+v} \leq \tilde{K}\ ||f||^{\tilde{S}(R)}_{g,a,r}. \tag{10.3}$$

It follows from proposition 4.3 that such numbers exist if N_o is sufficiently small. From (10.2) and (10.3) we derive the inequality

$$||\tilde{\Lambda}f||^{S_{N_o}(R)}_{g,a,r+v} \leq \sup_{s \in S_{N_o}(R)} |B(s)^{-1}|\{\tilde{K}\frac{F_{g,a,r+v}(|s-1|)}{F_{g,a,r+v}(|s|)} + |s|^{-v}\}\ ||f||^{\tilde{S}(R)}_{g,a,r},$$

where $F_{g,a,r+v}$ is defined by (9.38), which is valid for all $R \geq \tilde{R}_o$. Note that

$$\frac{F_{g,a,r+v}(|s-1|)}{F_{g,a,r+v}(|s|)} = 1 - \{ga|s|^{g-1} + (r+v)|s|^{-1}\} \cos \arg s + O(s^{g-2}). \tag{10.4}$$

If $d = 0$ we have

$$|B(s)^{-1}| = 1 + k\ \text{Re}\{\mu e^{i(k-1)\arg s}\}|s|^{k-1} + O(s^{k-1-1/p}), \tag{10.5}$$

and if $d < 0$,

$$|B(s)^{-1}| \leq C|s|^d,$$

where C is a positive constant.

We intend to prove the existence of positive numbers $\tilde{K}$ and $\tilde{R}_o$, independent of N_o, such that (10.3) holds for $N = N_o$ as well. Then it follows by induction on N that (10.3) is true for all $N \in \mathbb{N}$ and consequently,

$$||\tilde{\Lambda} f||^{\tilde{S}(R)}_{g,a,r+v} \leq \tilde{K} \; ||f||^{\tilde{S}(R)}_{g,a,r} .$$

In view of proposition 4.3 it is sufficient to consider the set $\{s \in S_{N_o}(R) : s \notin S(R)\}$, hence we may assume that $\arg s \leq \beta$.

I. Suppose that $d = 0$, $k < g$ and $a > 0$. In that case $v = 1 - \min\{g,\rho\}$ (ρ is the positive constant occurring in the definition of $S(R)$, cf. § 9.2, case III.1). From (10.4) and (10.5) we deduce that

$$|B(s)^{-1}| \frac{F_{g,a,r+v}(|s-1|)}{F_{g,a,r+v}(|s|)} \leq 1 - ga \cos\beta \; |s|^{g-1} + o(s^{g-1})$$

Now let $\tilde{K} > \max\{K, (ga \cos\beta)^{-1}\}$, where K is the positive number mentioned in proposition 4.3 (property 2). One easily verifies the existence of a number $\tilde{R}_o \geq R_o$, independent of N_o, such that (10.3) holds for $N = N_o$ and for all $R \geq R_o$.

II. Next, suppose that $d = 0$ and $k \geq g$. Then $v = 1 - k$ (cf. §9.3, case I). Let

$$\delta = -\max_{\theta \in [0,\beta]} \mathrm{Re}\{\mu e^{i(k-1)\theta}\}.$$

It follows from (10.1) and the third assumption of proposition 4.9 that $\delta > 0$. Using (10.4) and (10.5) we find

$$|B(s)^{-1}| \frac{F_{g,a,r+v}(|s-1|)}{F_{g,a,r+v}(|s|)} \leq 1 - ga \cos \arg s |s|^{g-1} - k\delta |s|^{k-1} + o(s^{k-1})$$

Assume that in addition to (4.6) the condition

$$(k,-\delta) < (g,a)$$

is satisfied. If $k > g$ we choose $\tilde{K} > \frac{1}{k\delta}$ and if $k = g$ we take $\tilde{K} > \max\{(\delta + a)^{-1}, (\delta + a \cos \beta)^{-1}\}$. In both cases there exists a positive number $\tilde{R}_o \geq R_o$, independent of N_o, such that (10.3) holds for $N = N_o$ and all $R \geq \tilde{R}_o$.

III. Finally, if $d < 0$, we have $v = d$. Let $\tilde{K} > C$. Then there exists a positive number $\tilde{R}_o$ with the required properties.

From the above discussion it can be easily deduced that, under the conditions mentioned there, the linear mappings $\widetilde{\Lambda}$ possess the properties stated in proposition 4.9. □

§11. Proof of proposition 4.16.

We shall limit the discussion to the study of sectors S of the form $S = S[\alpha_1,\pi]$, where $\alpha_1 \in (0,\pi)$. In all other cases the proof of proposition 4.16 either is strictly analogous to the one given below, or can be derived from it by means of the transformation described in §5. As we are mainly interested in the case that S 'is almost a half plane', we shall assume, moreover, that $\alpha_1 < \frac{\pi}{2}$. The assumptions (i) and (ii) of proposition 4.16 imply

$$d \leq 0 \text{ for all } d \in d(A) \tag{11.1}$$

and

$$\mathrm{Re}(e^{ik\pi}\mu) > 0 \text{ for all } k \in k(A) \text{ such that } k \geq k_i \text{ and all } \mu \in \mu_k(A). \tag{11.2}$$

We begin by defining a convenient set of S-proper regions. To this end we choose a $\theta \in (\alpha_1 + \frac{\pi}{2}, \pi)$ with the property that

$$\cos\{\arg \mu + (k-1)\theta\} < 0$$

for all $k \in k(A)$ such that $k \geq k_i$ and all $\mu \in \mu_k(A)$ (the existence of such a number θ follows from (11.2)).

Next we take a $\theta' \in (\theta,\pi)$. Let $R > 0$. By L we denote the line through $R\,e^{i(\alpha_1 + \frac{\pi}{2})}$ with directional angle α_1 and by H the closed half plane bounded from below by L. We choose a point s_o on L such that $\theta' \leq \arg s_o < \pi$. Let $H_1 = \{\zeta \in ¢: \mathrm{Im}\,\zeta \geq \mathrm{Im}\,s_o\}$. We now define S(R) as the intersection of H and H_1 (cf. fig. 3). Then $\{S(R), R > 0\}$ clearly is a set of S-proper regions.

As in the proof of proposition 4.3, we may assume that A is in canonical form. Let B be one of the diagonal blocks of A, of the form (9.1), and let Δ_B be the corresponding (left) difference operator. Further, let $g \in [k_{i-1}, k_i]$, $a \in \mathbb{R}^-$, $r \in \mathbb{R}$ and $R \in (0,\infty)$. We have to prove the existence of right inverses $\Lambda^{S(R)}_{g,a,r}$ of Δ_B possessing the properties mentioned in proposition 4.12. For the case that $d < 0$ this has been done in §9.4 already. (Indeed, we have not imposed any conditions there on the shape of the regions S(R), apart from the requirement that they should be S-proper regions). Now suppose that $d = 0$. We define a linear mapping Λ on $B_{g,a,r}(S(R))$ by the formula

$$\Lambda f(s) = -\sum_{h=o}^{\infty} Y_o(s) Y_o(s-h-1)^{-1} f(s-h), \qquad f \in B_{g,a,r}(S(R)), \tag{11.3}$$

where

$$Y_o(s) = e^{q(s)} s^G .$$

q(s) and G have been defined in (9.34). Let k be the rational number determined by the relation: $k(B) = \{k\}$ (or, equivalently, $k = \text{degr } \underline{q}$), and let $\mu = \mu_k$. We distinguish the following two cases according to whether $k \geq k_i$ or $k \leq k_{i-1}$,

case I $(k \geq k_i)$: $k \neq 0$ and $0 \leq (g,-a) < (k,c)$,

case II $(k \leq k_{i-1})$, which is subdivided into

II.1 : $(g,a) < (k,-b) \leq 0$ and II.2 : $g = k = 0$, $r < r_o$.

Here c,b and r_o are real numbers, to be specified later on.

We divide the regions S(R) into two parts, as follows. Let

$$S_1(R) = \{s \in S(R) : \arg s \geq \theta\},$$

and

$$S_2(R) = \{s \in S(R) : \arg s < \theta\}.$$

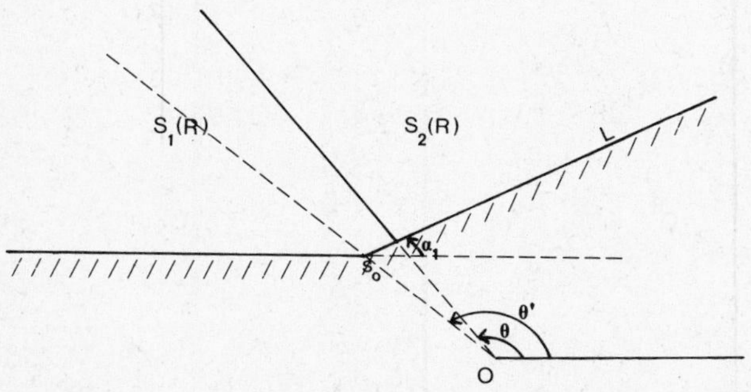

Fig. 3 $S(R) = S_1(R) \cup S_2(R)$.

(i) Suppose that $s \in S_1(R)$. In this case the infinite sum in (11.3) can be estimated by similar techniques as used in the proof of lemma 8.12 , as we shall show now. The capital K will be used to denote any constant that is independent of s, h and R, and of the choice of the function f. One easily verifies that

$$|Y_o(s-h)Y_o(s-h-1)^{-1}| \leq K, \qquad s \in S(1),\ h \in \mathbb{N}. \tag{11.4}$$

With the aid of an inequality similar to (9.41) we further find

$$|Y_o(s)Y_o(s-h)^{-1}| \leq |e^{q(s)-q(s-h)}|\,|\tfrac{s}{s-h}|^{\mathrm{Re}\gamma}|P(\log|\tfrac{s}{s-h}|)|, \qquad s \in S(1),\ h \in \mathbb{N} \cup \{0\}.$$

If we insert this into (11.3), taking into account the fact that $f \in B_{g,a,r}(S(R))$, using (11.4) and putting

$$\mathrm{Re}\ q(s) - a|s|^g = \Phi(s),$$

we obtain

$$|\Lambda f(s)| \leq K\ ||f||e^{a|s|^g}|s|^r \sum_{h=o}^{\infty} e^{\Phi(s)-\Phi(s-h)}|\tfrac{s-h}{s}|^{r-\mathrm{Re}\gamma}|P(\log|\tfrac{s}{s-h}|)|. \tag{11.5}$$

Let $\nu \in \mathbb{R}$ and $x \in (0,\infty)$. As $\theta > \frac{\pi}{2}$ we know that $|s-x| > |s|$, hence

$$|\tfrac{s-x}{s}|^{\nu} \leq 1 \qquad \text{if } \nu \leq 0. \tag{11.6}$$

On the other hand, if $\nu > 0$, we have

$$|\tfrac{s-x}{s}|^{\nu} \leq 2^{\nu} \qquad \text{if } x \in (0,|s|), \tag{11.7}$$

and

$$|\tfrac{s-x}{s}|^{\nu} \leq (x+1)^{\nu} \qquad \text{if } |s| \geq 1,\ x \in (0,\infty). \tag{11.8}$$

By the mean value theorem there exists a $\xi \in (0,x)$ such that

$$\frac{\Phi(s)-\Phi(s-x)}{x} = \mathrm{Re}\ q'(s-\xi) - ga|s-\xi|^{g-1}\cos\{\arg(s-\xi)\}.$$

As $\cos\{\arg(s-\xi)\} \leq \cos\theta$ and a is supposed to be a negative number, this implies

$$\frac{\Phi(s)-\Phi(s-x)}{x} \leq \mathrm{Re}\ q'(s-\xi) - ga|s-\xi|^{g-1}\cos\theta. \tag{11.9}$$

From (9.34) we deduce that

$$\mathrm{Re}\ q'(s-\xi) \leq k|s-\xi|^{k-1}[|\mu|\cos\{\arg\mu + (k-1)\arg(s-\xi)\} + C|s-\xi|^{-1/p}], \tag{11.10}$$

where C is a positive number independent of s and ξ. As a typical example showing how the infinite sum on the right-hand side of (11.5) can be estimated, let us consider the following case. Assume that $0 < k = g < 1$ and that $a < |\mu|(\cos\theta)^{-1}$. Let

$$|\mu| - a\cos\theta = -\delta.$$

Suppose that $h < |s|$. If $|s| \geq (\frac{2C}{\delta})^p$, then it follows from (11.9) and (11.10) that

$$\Phi(s) - \Phi(s-h) \leq -k|s-h|^{k-1}\frac{\delta}{2}h \leq -k\,2^{k-2}|s|^{k-1}\delta h. \tag{11.11}$$

If, on the other hand, $|s| \leq (\frac{2C}{\delta})^p$ we find

$$\Phi(s) - \Phi(s-h) \leq -k|s-h|^{k-1}\delta h + kC|s|^{k-1/p} \leq -k\,2^{k-1}|s|^{k-1}\delta h + kC(\frac{2C}{\delta})^{pk-1}. \tag{11.12}$$

Let Σ_2 denote the part of the infinite sum in (11.5) consisting of all terms with $h > |s|$ and let Σ_1 denote the sum of the remaining terms. Further, we put

$$k\,2^{k-2}\delta = \tilde{\delta}.$$

Utilizing (11.6),(11.7),(11.11) and (11.12) we obtain

$$\Sigma_1 \leq K\sum_{h=o}^{\infty} e^{-\tilde{\delta}|s|^{k-1}h} = K(1-e^{-\tilde{\delta}|s|^{k-1}})^{-1}$$

Since

$$(1-e^{-\tilde{\delta}|s|^{k-1}})^{-1} = \tilde{\delta}^{-1}|s|^{1-k}\{1+O(|s|^{k-1})\}$$

we conclude that

$$\Sigma_1 \leq K|s|^{1-k}. \tag{11.13}$$

Now suppose that $h > |s|$. Observing that

$$|q(s) - \mu s^k| \leq C'|s|^{k-1/p},$$

where C' is a positive number independent of s, and applying the mean value theorem to the function $\mathrm{Re}(\mu s^k) - a|s|^k$ instead of Φ, we find

$$\Phi(s) - \Phi(s-h) \leq -k|s-h|^{k-1}\delta h + C'(|s|^{k-1/p} + |s-h|^{k-1/p}),$$

and hence

$$\Phi(s) - \Phi(s-h) \leq -k\,2^{k-1}\delta\, h^k + C'(1+2^{k-1/p})\,h^{k-1/p}. \tag{11.14}$$

Utilizing (11.6), (11.8) and (11.14) and putting

$$k\,2^{k-1}\delta = \tilde{\delta}\,, \quad C'(1+2^{k-1/p}) = \tilde{C}\,,$$

we conclude that

$$\Sigma_2 \leq \sum_{h=o}^{\infty} e^{-\tilde{\delta}h^k + \tilde{C}h^{k-1/p}}(h+1)^{|r-\mathrm{Re}\gamma|}\tilde{P}\{\log(h+1)\} \leq K, \tag{11.15}$$

provided $R \geq 1$. $\tilde{P}$ is a polynomial with positive coefficients. Combining (11.5) with (11.13) and (11.15) we obtain

$$\|\Lambda f\|^{S_1(R)}_{k,a,r+1-k} \leq K \, \|f\|, \qquad R \geq 1,$$

provided

$$a < |\mu| (\cos \theta)^{-1}.$$

By means of similar arguments one derives the following estimates,

$$\|\Lambda f\|^{S_1(R)}_{g,a,r+1-g} \leq K \, \|f\| \quad \text{ if } g > k,\ a < 0,\ R \geq 1,$$

and, in the case that $k \geq k_i$,

$$\|\Lambda f\|^{S_1(R)}_{g,a,r+1-k} \leq K \, \|f\| \quad \text{ if } k \neq 0,\ g \leq k,\ a < 0,\ R \geq 1 \text{ (case I).}$$

In the derivation of the last estimate, essential use is made of the fact that, for all $s \in S_1(R)$ and all $\xi \geq 0$,

$$\cos\{\arg \mu + (k-1)\arg(s-\xi)\} \leq \max_{\beta \in [\theta,\pi]} \cos\{\arg \mu + (k-1)\beta\} < 0.$$

(ii) Next, suppose that $s \in S_2(R)$. With the aid of Cauchy's theorem it can be verified that $\Lambda f(s)$ may be represented by the formula

$$\Lambda f(s) = -\int_{C(s)} Y_o(s) Y_o(\zeta-1)^{-1}\{1 - e^{-2\pi i(s-\zeta)}\}^{-1} f(\zeta)d\zeta - B(s)^{-1} f(s), \tag{11.16}$$

where $C(s)$ is an unbounded closed contour defined in the following manner. Let

$$s' = s - \tfrac{1}{2}.$$

Let s_1 be the point on the boundary of $S(R)$ such that

$$\arg(s' - s_1) = \alpha_1.$$

$C_1(s)$ and $\ell(s)$ will denote the half-line from s_1 to infinity parallel to the real axis, and the segment between s_1 and s', respectively. Further, we define a path $C_2(s)$ in the same manner as the path $C(s)$ in the corresponding case (i.e. case I.b(i) or II.1) in §9.3 . Thus, in case I, $C_2(s)$ can be represented as follows,

$$C_2(s) = \{\zeta \in \mathbb{C} : \zeta^k = s'^k + e^{ik\alpha}x, \quad x \in (0,\infty)\},$$

where α has been chosen in such a way that $\alpha \in [\alpha_1, \pi]$ and

$$\operatorname{Re}(e^{ik\alpha}\mu) = \max_{\beta \in [\alpha_1,\pi]} \operatorname{Re}(e^{ik\beta}\mu).$$

In case II.1 on the other hand, $C_2(s)$ is the half-line from s' to infinity with directional angle $\arg s'$.

We now define $C(s)$ as the contour composed of $C_1(s)$, $\ell(s)$ and $C_2(s)$. It is readily verified that $C(s) \subset S(R)$ for all $s \in S_2(R)$. The contributions to the integral in (11.16) originating from $C_1(s), \ell(s)$ and $C_2(s)$ will be denoted by $I_1(s), I_\ell(s)$ and $I_2(s)$, respectively. The integral $I_\ell(s)$ has the same form as $I_2(s)$ in §9.3, the only difference being that, in the case considered here, the path of integration is bounded. By the same method that we used to estimate $I_2(s)$ in §9.3, case 2 (cf. p. 45, with $j = 2$, $\hat{\alpha}_2 = \alpha_1 + \pi$), one can prove that

$$\|I_\ell\|^{S_2(R)}_{g,a,r} \leq K \|f\|, \qquad R \geq 1, \tag{11.17}$$

provided that $g < 1$ and $a < 0$, or $g = 1$ and

$$0 < -a < \mathrm{Re}\{e^{i(\alpha_1+\pi)}(\mu_1 + 2\pi i)\}.$$

Here, the determinations of μ_1, $\arg \mu_1$ and $\arg(\mu_1 + 2\pi i)$ are to be chosen in such a way that $\arg(\mu_1 + 2\pi i) < \pi \leq \arg \mu_1 < \frac{3}{2}\pi$. The integral $I_2(s)$ resembles $I(s)$ in (9.48) but for the factor $\{1 - e^{-2\pi i(s-\zeta)}\}^{-1}$. From the definition of the path of integration $C_2(s)$ and the fact that $0 < \alpha \leq \arg s' \leq \theta' < \pi$, it follows easily that along this path the factor in question is bounded by a constant independent of s. In the same manner as in cases I.b(i) and II-1 in §9.3 (cf. pp. 48, 50, 51), we obtain the following estimates,

$$\|I_2\|^{S_2(R)}_{g,a,r+1-k} \leq K \|f\|, \qquad R \geq 1,$$

if $g < k$ and $a < 0$, or $g = k$ and $0 < -a < \max\limits_{\beta \in [\alpha_1, \pi]} \mathrm{Re}(e^{ik\beta}\mu)$ (case I), and

$$\|I_2\|^{S_2(R)}_{g,a,r+1-g} \leq K \|f\|, \qquad R \geq 1,$$

if $g > k$ and $a < 0$, or $g = k$ and $a < \min\{0, \min\limits_{\beta \in [\alpha_1, \theta']} \mathrm{Re}(e^{ik\beta}\mu)\}$ (case II.1).

Finally, let us consider the integral $I_1(s)$. This may be written in the form

$$I_1(s) = \int_0^\infty Y_0(s) Y_0(s_1 - x)^{-1}\{1 - e^{-2\pi i(s-s_1+x)}\}^{-1} f(s_1 - x)dx, \quad s \in S_2(R). \tag{11.18}$$

It is apparent from the definitions of s_1 and $S_2(R)$ that there exists a positive number ε such that, for all $s \in S_2(R)$ and all $x \in \mathbb{R}$, we have

$$\mathrm{Im}(s + x - s_1) \geq R\,\varepsilon.$$

Hence we deduce that

$$|1 - e^{-2\pi i(s-s_1+x)}|^{-1} \leq K e^{-2\pi\,\mathrm{Im}(s-s_1)}, \qquad s \in S_2(R),\ x \in \mathbb{R}. \tag{11.19}$$

(Here the constant K is not only independent of s, but also of x). With the aid of estimates similar to those used in the derivation of (11.17) one can prove that, under the same assumptions as made there (cf. first two lines after (11.17)),

$$|Y_o(s)Y_o(s_1)^{-1}e^{-2\pi i(s-s_1)}|e^{a(|s_1|^g-|s|^g)} \leq K, \qquad s\in S_2(R). \tag{11.20}$$

From (11.18), (11.19) and (11.20) we deduce the inequality

$$|I_1(s)|e^{-a|s|^g} \leq K\, e^{-a|s_1|^g} \int_0^\infty |Y_o(s_1)Y_o(s_1-x)^{-1}\, f(s_1-x)|dx, \qquad s\in S_2(R).$$

Note that $s_1\in S_1(R)$. The integral on the right-hand side of the inequality above can be estimated in very much the same way as the infinite sum in (i). One thus obtains estimates of the same form as those found in (i) with Λf replaced by I_1.

Collecting the results found in (i) and (ii), we conclude that

$$\|\Lambda f\|^{S(R)}_{g,a,r+1-k} \leq K\,\|f\|, \qquad R\geq 1,$$

if $g<k$ and $a<0$, or $g=k$ and $0<-a<c$ (case I), where

$$c=\min\{\mathrm{Re}[e^{i(\alpha_1+\pi)}(\mu_1+2\pi i)],\ \max_{\beta\in[\alpha_1,\pi]}\mathrm{Re}(e^{ik\beta}\mu)\}.$$

(Note that $c\to\min\{-\mathrm{Re}\,\mu_1,\ \max_{\beta\in[o,\pi]}\mathrm{Re}(e^{ik\beta}\mu)\}$ when $\alpha_1\to 0$). Furthermore we find that

$$\|\Lambda f\|^{S(R)}_{g,a,r+1-g} \leq K\,\|f\|, \qquad R\geq 1,$$

if $g>k$ and $a<0$, or $g=k$ and $a<-|\mu|\,|\cos\theta|^{-1}$ (case II.1).
Obviously, the last condition may be replaced by $a<-|\mu|$, provided we choose θ in such a way that $\cos\theta > \frac{|\mu|}{a}$. This implies that the definition of the regions S(R) becomes dependent of the value of a.

The case that $g=k=0$ (case II.2) is very simple. With the aid of an inequality proved by Birkhoff (cf. [4], p. 248) one obtains the following result:

$$\|\Lambda f\|^{S(R)}_{0,0,r+1} \leq K\,\|f\|, \qquad R\geq 1$$

provided $r<\mathrm{Re}\,\gamma-1$.

It follows immediately from the definition (11.3) and the estimates found above, that in all cases considered, the mapping Λ is an inverse of Δ_B possessing the properties mentioned in the proposition.

§12. Proofs of propositions 4.11. and 4.12.

§12.1 Proof of proposition 4.11.

Let $S = S[\frac{\pi}{2}, \frac{3\pi}{2}]$. We define a set of S-proper regions in the following manner. Let $\theta \in [-\pi, -\frac{\pi}{2})$ and let $R \geq 1$. By $\sigma(R)$ we denote the set

$$\sigma(R) = \{\zeta \in S : \operatorname{Re}[\zeta \log(e^{i\theta}\zeta)] = K_R\}, \tag{12.1}$$

where K_R is a nonpositive constant which is chosen in such a way that

$$d(\sigma(R), 0) = R.$$

It is not difficult to see that the points of this set form an unbounded continuous curve. A sketch of this curve, for $\theta = -\frac{3}{4}\pi$, $K_R = -2 \log 2$, is shown in fig. 4. For a more detailed description we refer the reader to the discussion of the paths $C(s)$ below. In particular it can be inferred from this discussion that the closed regions $S(R)$ consisting of all points lying to the left of or on the curve $\sigma(R)$ form a set of S-proper regions.

We may assume that the matrix function A is in canonical form. Let B be one of the diagonal blocks of A, of the form (9.1) and let Δ_B be the corresponding left difference operator. The homogeneous linear difference equation $\Delta_B\, y = 0$ has a fundamental matrix Y_o of the form

$$Y_o(s) = s^{ds}\, e^{q(s)}\, s^G,$$

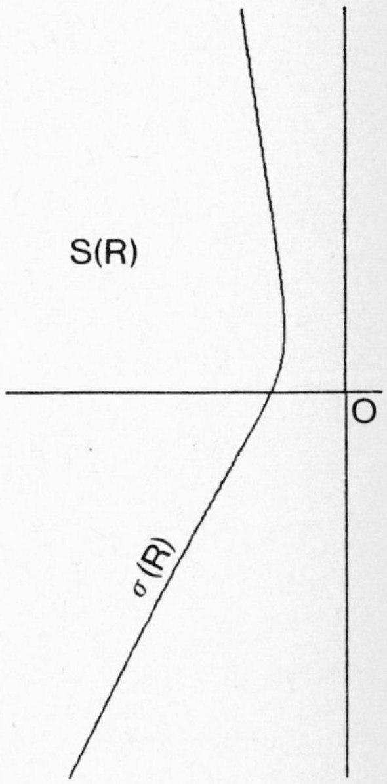

fig. 4

where d is a real number, and q and G are defined as in (9.34). By assumption, $d \neq 0$. The case that $d < 0$ has been treated in §9.4. Suppose, therefore, that $d > 0$.

Let $g \in [0,1]$, $a \in \mathbb{R}$, $r \in \mathbb{R}$ and let $R > 1$. We define a linear mapping Λ on $B_{g,a,r}(S(R))$ as follows

$$\Lambda f(s) = -\int_{C(s)} Y_o(s)Y_o(\zeta - 1)^{-1}\{1 - e^{2\pi i(s-\zeta)}\}^{-1} f(\zeta)d\zeta - B(s)^{-1}f(s),$$

$$f \in B_{g,a,r}(S(R)), \qquad s \in S(R), \tag{12.2}$$

where C(s) is the contour represented by the equation

$$C(s) = \{\zeta \in ₵ : \operatorname{Re}[\zeta \log(e^{i\theta}\zeta)] = \operatorname{Re}[s'\log(e^{i\theta}s')], \arg \zeta \in [\tfrac{\pi}{2}, \tfrac{3\pi}{2}]\},$$

to be described in upward direction. Here and below we use the abbreviation

$$s - \tfrac{1}{2} = s'.$$

First of all, let us investigate the shape of the contours C(s). Let

$$R(s) = \operatorname{Re}[s' \log(e^{i\theta}s')].$$

We have

$$\operatorname{Re}[\zeta \log(e^{i\theta}\zeta)] = \operatorname{Re} \zeta \log|\zeta| - \operatorname{Im} \zeta(\arg \zeta + \theta).$$

A simple argument shows that, for all $x \in (-\infty,\infty)$ the equation

$$\rho \log\sqrt{(\rho^2 + x^2)} - x(\operatorname{arctg} \frac{x}{\rho} + \pi + \theta) = R(s) \tag{12.3}$$

has exactly one solution $\rho \in (-\infty, 0)$. Thus we can parametrize the curve C(s) by putting

$$\zeta = \rho(x) + ix, \qquad x \in (-\infty,\infty).$$

It follows from (12.3) that the function $\frac{\rho(x)}{x} \log\sqrt{(\rho(x)^2 + x^2)}$ is bounded on $(1,\infty)$ and on $(-\infty,-1)$. Hence we deduce that

$$\rho(x) = O(\frac{|x|}{\log|x|}) \quad \text{as } |x| \to \infty.$$

Differentiating (12.3) with respect to x we obtain

$$\rho'(x) = (\operatorname{arctg} \frac{x}{\rho(x)} + \pi + \theta)\{1 + \log\sqrt{(\rho(x)^2 + x^2)}\}^{-1}, \qquad x \in (-\infty,\infty). \tag{12.4}$$

We now proceed to derive a suitable estimate for the integral on the right-hand side of (12.2). As before, a capital K will denote any constant independent of s, R and the choice of f. From the definition of Y_o we derive the inequalities

$$|Y_o(s)Y_o(s')^{-1}| \leq K\ |s|^{d/2}, \qquad s \in S(1), \tag{12.5}$$

and

$$|Y_o(\zeta)Y_o(\zeta - 1)^{-1}| \leq K\ |\zeta|^{d}, \qquad \zeta \in S(1). \tag{12.6}$$

In the latter inequality the constant K is independent of ζ. Furthermore, we find

$$|Y_o(s')Y_o(\zeta)^{-1}| \leq |\ e^{d(s' \log s' - \zeta \log \zeta) + q(s') - q(\zeta)}\ (\frac{s'}{\zeta})^{G}|,$$

$$s \in S(1),\ \zeta \in S(1).$$

Recalling the definition of C(s), we conclude that

$$|Y_o(s')Y_o(\zeta)^{-1}| \leq e^{\mathrm{Re}\{-id\theta(s'-\zeta) + q(s') - q(\zeta)\}}\ |(\frac{s'}{\zeta})^{G}|, \ \zeta \in C(s),$$

$$s \in S(R),\ R \geq 1, \tag{12.7}$$

Let

$$s' = \rho(x_o) + i\,x_o.$$

For all $\zeta \in C(s)$ we have

$$1 - e^{2\pi i(s-\zeta)} = 1 + e^{2\pi i\{\rho(x_o) - \rho(x)\}}e^{-2\pi(x_o - x)}.$$

It follows from (12.4) that

$$|\rho'(x)| < \frac{\pi}{1 + \log R}, \qquad (R \geq 1) \tag{12.8}$$

for all $x \in (-\infty, \infty)$. Hence we deduce that

$$|1 - e^{2\pi i(s-\zeta)}|^{-1} \leq K, \qquad s \in S(1),\ \zeta \in C(s). \tag{12.9}$$

If Im $\zeta \geq$ Im s we can improve this estimate and get

$$|1 - e^{2\pi i(s-\zeta)}|^{-1} \leq K\ e^{-2\pi(x-x_o)}, \quad s \in S(1),\ x \in (x_o, \infty). \tag{12.10}$$

We split the path of integration C(s) into two parts, one part containing all values of ζ such that Im $\zeta \geq$ Im s and the other containing the values of ζ for which Im $\zeta \leq$ Im s. The corresponding contributions to the integral in (12.2) will be denoted by $I^+(s)$ and $I^-(s)$, respectively. We define two real functions u and φ_2 on $(-\infty,\infty)$ as follows,

$$u(x) = |\rho(x) + ix|, \qquad x \in (-\infty,\infty),$$

and

$$\varphi_2(x) = (\mathrm{Re}\gamma - r - d)\log u(x), \qquad x \in (-\infty,\infty).$$

Further, we define a function φ_1^+ on (x_0,∞) by

$$\varphi_1^+(x) = (2\pi + d\theta)x + \mathrm{Re}\{q(\rho(x) + ix)\} - a\,u(x)^g, \qquad x \in (x_0,\infty),$$

and a function φ_1^- on $(-\infty,x_0)$ by

$$\varphi_1^-(x) = d\theta\, x + \mathrm{Re}\,\{q(\rho(x) + ix)\} - a\,u(x)^g, \qquad x \in (-\infty,x_0).$$

One easily verifies that the estimates (9.40) and (9.41) are again valid. Utilizing (9.40), (9.41), (12.5) - (12.7), (12.9) and (12.10), and putting

$$\varphi_1^\pm(x) + \varphi_2(x) = \varphi^\pm(x), \qquad \pm(x - x_0) \in (0,\infty),$$

and

$$\|f\|_{g,a,r}^{S(R)} = \|f\|,$$

we obtain the following inequalities

$$\frac{|I^+(s)|}{F_{g,a,r+3d/2}(|s|)} \leq K\ \|f\| \int_{x_0}^{\infty} e^{\varphi^+(x_0) - \varphi^+(x)}\, |P\{\log(\frac{u(x_0)}{u(x)})\}|\,|\rho'(x) + i|\,dx, \tag{12.11}$$

and

$$\frac{|I^-(s)|}{F_{g,a,r+3d/2}(|s|)} \leq K\ \|f\| \int_{-\infty}^{x_0} e^{\varphi^-(x_0) - \varphi^-(x)}\, |P\{\log(\frac{u(x_0)}{u(x)})\}|\,|\rho'(x) + i\,|\,dx \tag{12.12}$$

(cf. (9.38) for the definition of $F_{g,a,r+3d/2}$).

Obviously $u(x) \geq 1$. Differentiating u we find

$$u'(x) = \frac{\rho(x)\rho'(x) + x}{\sqrt{(\rho(x)^2 + x^2)}}, \qquad x \in (-\infty,\infty).$$

Hence

$$|u'(x)| \leq |\rho'(x)| + 1.$$

In view of (12.8) this implies

$$|u'(x)| \leq 1 + \pi(1 + \log R)^{-1}, \qquad x \in (-\infty,\infty). \tag{12.13}$$

From (12.8) we deduce that

$$|\rho'(x) + i| \leq \pi + 1, \qquad x \in (-\infty,\infty). \tag{12.14}$$

It is easily seen that φ_2 and φ_2' satisfy inequalities of the form (8.5) and (8.6), respectively. The derivatives of φ_1^+ and φ_1^- are given by the expressions

$$(\varphi_1^+)'(x) = 2\pi + d\theta - \mathrm{Im}\ \mu_1 - g\,a\,u(x)^{g-1}u'(x) + R(x), \qquad x \in (x_o,\infty),$$

and

$$(\varphi_1^-)'(x) = d\theta - \mathrm{Im}\ \mu_1 - g\,a\,u(x)^{g-1}u'(x) + R(x), \qquad x \in (-\infty,x_o),$$

where

$$R(x) = \mathrm{Re}[\mu_1\rho'(x) + \sum_{ph=1}^{p-1} h\,\mu_h\,\{\rho(x) + ix\}^{h-1}(\rho'(x) + i)], \qquad x \in (-\infty,\infty).$$

We may choose θ in such a way that $\mathrm{Im}\ \mu_1 \neq d\theta \bmod 2\pi$. Further, we take the determination of μ_1 so that

$$d\theta < \mathrm{Im}\ \mu_1 < d\theta + 2\pi.$$

Let

$$c = \min\{\mathrm{Im}\ \mu_1 - d\theta,\ d\theta + 2\pi - \mathrm{Im}\ \mu_1\}.$$

Assume that $(g,|a|) < (1,c)$ and choose a positive number δ such that $\delta < c$ if $g < 1$, or $\delta < c - |a|$ if $g = 1$. With the aid of (12.8) and (12.13) one easily verifies the existence of a positive number R_o such that

$$(\varphi_1^+)'(x) \geq \delta, \qquad x \in (x_o,\infty), \tag{12.15}$$

and

$$(\varphi_1^-)'(x) \leq -\delta, \qquad x \in (-\infty,x_o), \tag{12.16}$$

if $R \geq R_o$. Using (12.8),(12.11) - (12.16) and applying lemma 8.1, we obtain the following result

$$||\Lambda f||^{S(R)}_{g,a,r+3d/2} \le K\ ||f||^{S(R)}_{g,a,r} ,$$

provided that $(g,|a|) < (1,c)$ and $R \ge R_o$. Moreover, the constant R_o is independent of r.

Let $F(s,\zeta) = Y_o(s)Y_o(\zeta-1)^{-1}\{1-e^{2\pi i(s-\zeta)}\}^{-1}f(\zeta)$. From the properties of the paths C(s) and the exponential decrease of $F(s,\zeta)$ as $|\zeta| \to \infty$ on C(s), it follows that

$$\int_{C(s-1)} F(s,\zeta)d\zeta = \int_{C(s)-1} F(s,\zeta)d\zeta.$$

It is now easily verified that the mapping Λ is both a right and a left inverse of Δ_B, possessing the properties mentioned in proposition 4.11.

REMARK. If instead of $s' = s-\frac{1}{2}$ we would take $s' = s-\varepsilon$ with $\varepsilon \in (0,\frac{1}{2})$ we would find that $\Lambda f \in B_{g,a,r+d(1+\varepsilon)}(S(R))$. (Although the paths C(s) would be different from those introduced above we would obtain the same mappings Λ). Hence we conclude that $\Lambda f \in B_{g,a,r+d'}(S(R))$ for all $d' > d$.

§12.2 Proof of proposition 4.12.

Note that (4.13) is equivalent to the following condition: for all $k \in k(A)$ such that $k \ge k_i$ and every $\mu \in \mu_k(A)$ there exists a real number $\alpha \in (\max\{\alpha_1,\alpha_2-\frac{\pi}{k}\}, \min\{\alpha_1+\frac{\pi}{k},\alpha_2\})$ such that

$$\mathrm{Re}(\mu e^{ik\alpha}) > 0.$$

If $A^{-1} \in \mathrm{End}(n;A_o(S))$ we define the regions S(R) in the same manner as in §9.2. In that case the conclusions of proposition 4.12 follow immediately from the observations made in the proof of proposition 4.3, cf. in particular the discussions of the integrals I(s), cases I.a), I.b)(i) , II.1 and II.2 in §9.3, and of the case $d < 0$ in §9.4

If, on the other hand, $S = S[\frac{\pi}{2},\frac{3}{2}\pi]$ and $A^{-1} \notin \mathrm{End}(n;A_o(S))$, we define the regions S(R) as in §12.1. Let $k \in k(A)$ such that $k \ge k_i$. Corresponding to each $\mu \in \mu_k(A)$ we choose a real number $\alpha \in (\frac{\pi}{2},\frac{3}{2}\pi)$ such that $\mathrm{Re}(\mu e^{ik\alpha}) > 0$. Let $\alpha_1(k)$ be the smallest of these numbers and $\alpha_2(k)$ the largest. Now, (12.8) implies that the slope of the tangent to $\sigma(R)$ (the boundary of S(R)) in any point of this curve, is larger in absolute value than $\frac{1}{\pi}(1+\log R)$. Hence, if we take R so large that

$$\frac{\pi}{2} - \operatorname{arctg}\{\frac{1}{\pi}(1 + \log R)\} < k \min\{\alpha_1(k) - \frac{\pi}{2}, \frac{3}{2}\pi - \alpha_2(k)\}, \tag{12.17}$$

then it follows that all half-lines starting from a point in $\chi_k(S(R))$ with directional angle $\alpha \in [\alpha_1(k), \alpha_2(k)]$ are contained in $\chi_k(S(R))$.

As before, we assume that A is in canonical form and consider a block B of the form (9.1). For the case that $d < 0$ the statements of the proposition have already been proved in §9.4 and for the case that $d > 0$ in §12.1. Now suppose that $d = 0$. Let R be taken so large that (12.17) holds for all $k \in k(A)$ such that $k \geq k_i$. It is easily verified that we can construct mappings Λ on $B_{g,a,r}(S(R))$ under similar conditions and in a similar manner as described in §9.3. As regards the discussion of the integrals I(s), we need only consider the cases I.a), I.b) (i), II.1 and II.2. It follows from the results obtained there that Λ has the required properties. This completes the proof of proposition 4.12.

CHAPTER III. *NONLINEAR DIFFERENCE EQUATIONS.*

§13. *Introduction.*

This chapter is concerned with systems of nonlinear difference equations of the form

$$y(s-1) = \varphi(s,y(s)) \tag{13.1}$$

or

$$y(s+1) = \varphi(s,y(s)), \tag{13.2}$$

where s is a complex variable, y and φ are n-dimensional vector functions, $n \in \mathbb{N}$.

A typical example of the kind of problem we shall study is the following. Assume that:

(i) φ is holomorphic on the set $S \times U$, where S is an open sector and U is a neighbourhood of $y_o \in \mathbb{C}^n$.

(ii) φ can be represented asymptotically by a series of the form

$$\hat{\varphi}(s,y) = \sum_{h=h_o}^{\infty} \tilde{\varphi}_h(y)s^{-h/p} \tag{13.3}$$

as $|s| \to \infty$ in S, where h_o is a nonpositive integer, and, moreover, this asymptotic expansion is uniformly valid on all sets $S' \times U$, where S' is a closed subsector of S. Under these conditions, the functions $\tilde{\varphi}_h$ are holomorphic on U for all $h \geq h_o$. Hence the following, and final, assumption makes sense,

(iii) the formal equation

$$y(s \pm 1) = \hat{\varphi}(s,y(s))$$

has a solution $\tilde{f} \in (\mathbb{C}[[s^{-1/p}]])^n$ with the property that $\tilde{f}(\infty) = y_o$. Here the lower sign is used when S is a left sector, the upper sign when S is a right sector. We shall call $\tilde{f}$ a formal solution of (13.1) or (13.2)

We are interested in holomorphic solutions of (13.1) or (13.2) which can be represented asymptotically by $\tilde{f}$.

Systems of this type have been treated by Harris and Sibuya in [17] (cf. also [6], [16]), in the case that both matrices $D_2\varphi(\infty,y_o)$ and $D_2\varphi(\infty,y_o) - I$ are invertible. In this chapter we shall extend their results to a larger class of equations.

Apart from studying 'ordinary' asymptotics, we shall be concerned with equations where the function φ satisfies certain Gevrey conditions and investigate the existence of solutions belonging to a corresponding Gevrey class of holomorphic functions. The results presented in this chapter show some analogy to those obtained by Ramis for differential equations (cf.[30]) but are much less far-reaching. In particular, the concept of 'g-summability' of formal solutions appears to have only a very limited use in the case of difference equations.

In §17 the theory developed in this chapter is applied to block-diagonalization and -triangularization problems. The results are used to simplify linear systems of difference equations in §18.

§14. Preliminaries.

In this section we introduce some new definitions and prove two preparatory lemmas.

DEFINITION. *Let F be a decreasing set of closed unbounded regions* $\{G(R), R>0\}$ *of the complex plane such that* $d(G(R),0)\to\infty$ *as* $R\to\infty$ *and let* $y_o\in\mathbb{C}^n$. *By* $V_o(F,y_o)$ *we denote the set of complex-valued functions* φ[6] *with the following properties:*
there exist a positive number R *and a neighbourhood* U *of* y_o *such that*
(i) φ *is continuous on the set* $G(R)\times U$ *and holomorphic in its interior.*
(ii) φ *admits an asymptotic expansion* $\hat{\varphi}$ *of the form* (13.3) *as* $|s|\to\infty$ *in* G(R), *uniformly on* $G(R)\times U$.

We use the following notation:

$$R_N(\varphi;s,y)=\varphi(s,y)-\sum_{h=h_o}^{N-1}\tilde{\varphi}_h(y)s^{-h/p}, \qquad \text{if } N\in\mathbb{N},$$

and

$$R_o(\varphi;s,y)=\varphi(s,y).$$

If $y_o\in\mathbb{C}^n$ and $\delta>0$, then $U(y_o,\delta)$ will denote the set

$$U(y_o,\delta)=\{y\in\mathbb{C}^n : |y-y_o|<\delta\}.$$

The 'Gevrey version' of the preceding definition is as follows.

6) properly speaking: equivalence classes of these functions, cf. Chapter I, p. 9.

DEFINITION. *Let* F *and* y_o *be as in the preceding definition and let* g *be a positive number. By* $V_g(F,y_o)$ *we denote the set of functions* $\varphi \in V_o(F,y_o)$ *with the additional property that there exist positive constants* R, B *and* C *such that, for all* $N \in \mathbb{N}$,

$$\sup_{(s,y)\in G(R)\times U} |s^{N/P} R_N(\varphi;s,y)| \leq C(N!)^{1/pg} B^N$$

This definition also applies to closed sectors S as defined in §3 (pp. 11, 12). For open and half-open sectors we use the following definition.

DEFINITION. *Let* S *be a sector,* $y_o \in \mathbb{C}^n$ *and let* $g \geq 0$. *By* $V_g(S,y_o)$ *we denote the set*

$$V_g(S,y_o) = \bigcap_{\overline{S}' \subset S} V_g(\overline{S}',y_o).$$

Note that, in the last definition, the neighbourhood U of y_o on which φ is holomorphic may depend on the choice of S'.

LEMMA 14.1. *Let* F *be a decreasing set of closed unbounded regions* $\{G(R), R>0\}$ *of the complex plane such that* $d(G(R),0) \to \infty$ *as* $R \to \infty$, *let* $g>0$, $f_1 \in A_g(F)$ *and* $f_2 \in A_g(F)$. *Let* B, C_1, C_2 *and* R *be positive numbers, chosen in such a way that, for all nonnegative integers* N, *the inequalities*

$$\sup_{s\in G(R)} |s^{N/p} R_N(f_j;s)| \leq C_j(N!)^{1/pg} B^N, \qquad j = 1,2,$$

hold. There exists a positive constant K *which depends exclusively on* g, *such that, for all nonnegative integers* N *the following estimate is valid*

$$\sup_{s\in G(R)} |s^{N/p} R_N(f_1f_2;s)| \leq K\, C_1C_2(N!)^{1/pg} B^N.$$

PROOF: A simple calculation shows that the following identity holds for all $s \in G(R)$ and all nonnegative integers N,

$$R_N(f_1f_2;s) = R_o(f_1;s)R_N(f_2;s) + \sum_{j=1}^{N} R_j(f_1;s)\{R_{N-j}(f_2;s) - R_{N+1-j}(f_2;s)\}.$$

Multiplying both sides by $s^{N/p}$ and observing that

$$s^{(N-j)/p}\{R_{N-j}(f_2;s)-R_{N+1-j}(f_2;s)\} = \lim_{|s|\to\infty} s^{(N-j)/p} R_{N-j}(f_2;s),$$

we derive the inequality

$$\sup_{s \in G(R)} |s^{N/p} R_N(f_1 f_2;s)| \leq \sum_{j=o}^{N} \{\frac{j!(N-j)!}{N!}\}^{1/pg} C_1 C_2 (N!)^{1/pg} B^N.$$

Now consider the finite sum figuring on the right-hand side of this inequality. For all nonnegative integers N we have

$$\sum_{j=o}^{N} \{\frac{j!(N-j)!}{N!}\}^{1/pg} \leq 2 \sum_{j \leq N/2} \{\frac{j!}{(N-j+1)..N}\}^{1/pg}.$$

Clearly, $j \leq \frac{N}{2}$ implies that

$$\frac{j!}{(N-j+1)..N} \leq 2^{-j}.$$

Hence it follows that

$$\sum_{j=o}^{N} \{\frac{j!(N-j)!}{N!}\}^{1/pg} \leq 2 \sum_{j=o}^{\infty} 2^{-j/pg},$$

which proves the lemma. In particular we see that for K we may take any number such that

$$K \geq \frac{2}{1-2^{-1/pg}} . \tag{14.2}$$

LEMMA 14.3. *Let* $\varphi \in V_g(F,y_o)$, *where F is defined as in the preceding lemma,* $y_o \in \mathbb{C}^n$ *and* $g \geq 0$. *If* f *is an element of* $(A_g(F))^n$ *with the property that* $\hat{f}(\infty) = y_o$, *then the function* ψ *defined by*

$$\psi(s,z) = \varphi(s,z+f(s))$$

belongs to $V_g(F,0)$.

PROOF: We shall restrict the discussion to the case that $g > 0$. Without loss of generality we may assume that $y_o = 0$ and that $\hat{\varphi}(s,y)$ is of the form (13.3) with $h_o = 0$ (if $h_o < 0$ then consider the function $s^{h_o/p}\varphi$).

Let R_o and δ_o be positive numbers such that φ is holomorphic in int $G(R_o) \times U(0,\delta_o)$. On this set φ can be represented by a power series in y as follows,

$$\varphi(s,y) = \sum_{|m| \geq o} \varphi_m(s) y^m,$$ [7)]

where $m = (m_1,..,m_n)$, m_i is a nonnegative integer for $i = 1,..,n$. The functions φ_m are elements of $A_o(F)$, admitting asymptotic expansions of the form

$$\hat{\varphi}_m(s) = \sum_{h=o}^{\infty} \varphi_{hm} s^{-h/p},$$

7) If $y = (y_1,..,y_n)$ then $y^m = y_1^{m_1}..y_n^{m_n}$; $|m| = m_1 + .. + m_n$.

where $\varphi_{hm} \in \mathbb{C}$. Furthermore, the functions $\tilde{\varphi}_h$ occurring in (13.3) are holomorphic in $U(0,\delta_o)$ and can be represented by the power series

$$\tilde{\varphi}_h(y) = \sum_{|m| \geq 0} \varphi_{hm}\, y^m,$$

for all nonpositive integers h. Let $y = (y_1,..,y_n)$. Let δ be any positive number smaller than δ_o. By Cauchy's formula we have, for all $s \in G(R_o)$, all m such that $|m| \geq 0$ and all nonnegative integers N,

$$R_N(\varphi_m;s) = \frac{1}{(2\pi i)^n} \oint_{C_1} .. \oint_{C_2} \frac{R_N(\varphi;s,y)}{y_1^{m_1+1} .. y_n^{m_n+1}}\, dy_1..dy_n, \tag{14.4}$$

where C_j is the circle $|y_j| = \delta$ (described in counter-clockwise direction), $j = 1,..,n$. Since $\varphi \in V_g(F,0)$ there exist positive constants B and C such that, for all nonnegative integers N, the following inequality holds,

$$\sup_{(s,y) \in G(R_o) \times U(0,\delta_o)} |s^{N/p} R_N(\varphi;s,y)| \leq C(N!)^{1/pg} B^N.$$

Inserting this into (14.4) and letting δ approach δ_o, we obtain the estimate

$$\sup_{s \in G(R_o)} |s^{N/p} R_N(\varphi_m;s)| \leq C\, \delta_o^{-|m|} (N!)^{1/pg} B^N, \tag{14.5}$$

valid for all m such that $|m| \geq 0$ and all $N \in \mathbb{N} \cup \{0\}$.
Now, let $f = (f_1,..,f_n) \in (A_g(F))^n$ and suppose that $\hat{f}(\infty) = 0$. Then there exist positive numbers R_1, D and F such that f is holomorphic in int $G(R_1)$ and

$$\sup_{s \in G(R_1)} |s^{N/p} R_N(f_j;s)| \leq F(N!)^{1/pg} D^N$$

for all $j \in \{1,..,n\}$ and all $N \in \mathbb{N}$. Let K be a positive number satisfying (14.2) and let

$$\delta_1 < \frac{\delta_o}{2K}.$$

Since $\hat{f}(\infty) = 0$ there exists a number $R_2 \geq R_1$ such that

$$\sup_{s \in G(R_2)} |f(s)| \leq \delta_1.$$

Put

$$\max(D, \frac{FD}{2\delta_1}) = \tilde{D}.$$

For all $z=(z_1,..,z_n)\in U(0,\delta_1)$, all $j\in\{1,..,n\}$ and all $N\in\mathbb{N}\cup\{0\}$ we now have

$$\sup_{s\in G(R_2)} |s^{N/p} R_N(f_j+z_j;s)| \leq 2\delta_1 (N!)^{1/pg} \tilde{D}^N.$$

Application of lemma 14.1 yields the following inequality, which is valid for all $z\in U(0,\delta_1)$, all m such that $|m|\geq 0$ and all $N\in\mathbb{N}\cup\{0\}$,

$$\sup_{s\in G(R_2)} |s^{N/p} R_N((f+z)^m;s)| \leq K^{|m|-1} (2\delta_1)^{|m|} (N!)^{1/pg} \tilde{D}^N.$$

Combining this with (14.5) and once more applying lemma 14.1 we obtain, for all $N\in\mathbb{N}\cup\{0\}$, the estimate

$$\sup_{(s,z)\in G(R)\times U} |s^{N/p} R_N(\psi;s,z)| \leq \tilde{C} (N!)^{1/pg} \tilde{B}^N,$$

where

$$R=\max\{R_o,R_2\},\quad U=U(0,\delta_1),\quad \tilde{B}=\max\{B,\tilde{D}\}$$

and

$$\tilde{C}=C \sum_{|m|\geq o} \left(\frac{2\delta_1 K}{\delta_o}\right)^{|m|}.$$

Obviously, ψ is continuous on $G(R)\times U$ and holomorphic in its interior. Thus,

$$\psi\in V_g(F,0).$$

COROLLARY. *Let* $\varphi\in V_g(S,y_o)$, *where* S *is a sector,* $y_o\in\mathbb{C}^n$ *and* $g\geq 0$. *If* $f\in(A_g(S))^n$ *and* $\hat{f}(\infty)=y_o$, *then the function* ψ *defined in lemma* 14.3 *belongs to* $V_g(S,0)$.

REMARK. It follows from (14.5) that $\varphi_m\in A_g(F)$. More generally, one easily verifies that $\varphi\in V_g(F,y_o)$ implies: $\varphi_m\in A_g(F)[s^{1/p}]$ for all m such that $|m|\geq 0$, and, further: $D_2\varphi\in V_g(F,y_o)$.

§15. The main results.

THEOREM 15.1. *Let* $\varphi\in(V_g(S_o,y_o))^n$, *where* $S_o=S(\alpha_1^o,\alpha_2^o)$ *such that* $\pm S_o$ *is a right sector,* $y_o\in\mathbb{C}^n$ *and* $g\in[0,1]$. *Suppose that the equation*

$$y(s\pm 1)=\varphi(s,y) \tag{15.2}$$

has a formal solution $\tilde{f} \in (\mathbb{C}[\![s^{-1/p}]\!]_{pg})^n$ *with the property that* $\tilde{f}(\infty) = y_o$.
Set

$$D_2\hat{\varphi}(s,\tilde{f}) = \tilde{A}$$

and assume that

(i) $\tilde{A}^{-1} \in \mathrm{End}(n;\mathbb{C}[\![s^{-1/p}]\!])$,

(ii) *for all* $k \in k(\tilde{A})$ *such that* $k \geq g$ *and* $k(\alpha_2^o - \alpha_1^o) > \pi$ *the following condition is satisfied*

$$(\alpha_1^o, \alpha_2^o - \frac{\pi}{k}) \cap \Sigma_k^{\pm}(\tilde{A}) = \phi.$$

(iii) *if* $\alpha_2^o - \alpha_1^o < \pi$ *then* $[\alpha_2^o - \pi, \alpha_1^o]$ *contains at most one element of* $\Sigma_{1,j}^{\pm}(\tilde{A})$ *for each* $j \in \{1,..,m\}$ (cf. definition on p.8).

Then equation (15.2) *possesses a solution* $y \in (A_g(S_o))^n$ *with the property that* $\hat{y} = \tilde{f}$. *If, in addition,*

(iv) *for all* $k \in k(\tilde{A})$ *such that* $k \geq g$ *and* $0 < k(\alpha_2^o - \alpha_1^o) \leq \pi$ *the condition*

$$[\alpha_2^o - \frac{\pi}{k}, \alpha_1^o] \cap \Sigma_k^{\pm}(\tilde{A}) = \phi$$

is fulfilled, then the solution with these properties is unique.

<u>COROLLARY</u>. *Suppose that* $g \neq 0$ *and* $\alpha_2^o - \alpha_1^o > \frac{\pi}{g}$. *If the conditions of theorem* 15.1 *are satisfied, then the series* $\tilde{f}$ *is strongly* g-*summable in all directions* $\alpha \in (\alpha_1^o + \frac{\pi}{2g}, \alpha_2^o - \frac{\pi}{2g})$. (cf. [30]).

<u>PROOF OF THEOREM 15.1</u> : We shall assume that S_o is a left sector. The case that S_o is a right sector can be treated analogously. The proof will be given in several steps.

I. To begin with we prove the theorem under the simplifying assumption that $\tilde{f} = 0$. Let the vector function φ_o and matrix function A be defined by

$$\varphi_o(s) = \varphi(s,0) \quad , \quad A(s) = D_2\varphi(s,0).$$

By assumption, $\varphi \in (V_g(S_o,0))^n$ and $\hat{\varphi}(s,0) = 0$.
Hence it follows that

$$\varphi_o \in (A_{go}(S_o))^n.$$

Furthermore, $\varphi \in (V_g(S_o,0))^n$ implies that $\tilde{A} \in \mathrm{End}(n;\mathbb{C}[\![s^{-1/p}]\!]_{pg}[s^{1/p}])$. In view of condition (i) we have

$$\tilde{A}^{-1} \in \mathrm{End}(n;\mathbb{C}[\![s^{-1/p}]\!]) \cap \hat{M}_o.$$

Since $\hat{A} = \tilde{A}$, we conclude that

$$A^{-1} \in \mathrm{End}(n;A_o(S_o)) \cap M_o(S_o).$$

Let Δ denote the left difference operator corresponding to A and let ψ be the vector function defined by

$$\psi(s,y) = \varphi(s,y) - \varphi_o(s) - A(s)y.$$

Clearly, (15.2) is equivalent with

$$\Delta y(s) = \varphi_o(s) + \psi(s,y(s)). \tag{15.3}$$

We shall seek a vector function y with the property that

$$y(s) = \Lambda[\varphi_o(s) + \psi(s,y(s))],$$

where Λ is a right inverse of Δ defined on a suitable Banach space.

Let S be a left sector of the form $S = S[\alpha_1,\alpha_2]$ such that $S \subset S_o$. By increasing the aperture of S, if necessary, we can achieve that, for all $k \in k(A)$ such that $k \geq g$, $k \neq 0$, the following conditions hold,

$$(\alpha_1^o,\alpha_1] \cap \Sigma_k^-(A) = [\alpha_2 - \tfrac{\pi}{k}, \alpha_2^o - \tfrac{\pi}{k}) \cap \Sigma_k^-(A) = \phi. \tag{15.4}$$

Then the conditions of proposition 4.3 , or, if (iv) is satisfied, of proposition 4.12 are fulfilled (If $k(A) \cup \{0,1\} = \{k_o,..,k_\ell\}$ we choose the positive integer i in such a way that $k_{i-1} < g \leq k_i$. Then, $k \geq g$ implies $k \geq k_i$ and vice versa). Let $\{S(R), R>0\}$ be a set of S-proper regions as defined in §9.2 and let $v = v_i$ be the real number mentioned in proposition 4.3. From the fact that $\varphi_o \in (A_{go}(S_o))^n$ we deduce the existence of a negative number a and a positive number R_1 such that

$$\varphi_o|_{S(R_1)} \in B_{g,a,r-v}(S(R_1)), \tag{15.5}$$

for all $r \in \mathbb{R}$ (cf. proposition 3.1)

Observing that $D_2\psi(s,0) = 0$ one easily proves the following property of ψ (cf. [16],[40]): there exist positive numbers δ and R_2, a positive constant K independent of s, and a constant $r_1 \in \frac{1}{p}\mathbb{Z}$, such that, for all $y_1 \in U(0,\delta)$, all $y_2 \in U(0,\delta)$ and all $s \in S(R_2)$ the inequality

$$|\psi(s,y_1) - \psi(s,y_2)| \leq K\, |s|^{r_1} \max\{|y_1|,|y_2|\}\, |y_1 - y_2| \tag{15.6}$$

holds. Let $r_o' > \max(0, r_o - v)$, where r_o is the nonnegative constant mentioned in proposition 4.3, and let R_3 be a positive number such that

$$e^{aR_3^g} R_3^{-r_o'} \leq \delta.$$

Suppose that $R > \max(1, R_2, R_3)$ and $r \leq -r_0'$. By E we denote the unit sphere of $B_{g,a,r}(S(R))$.

In order to simplify the notation we put

$$\|\cdot\|^{S(R)}_{g,a,r} = \|\cdot\|.$$

Obviously, $y \in E$ implies that $y(s) \in U(0,\delta)$ for all $s \in S(R)$. Thus, if y_1 and y_2 are elements of E, then it follows from (15.6) that

$$|\psi(s,y_1(s)) - \psi(s,y_2(s))| \leq K\ |s|^{r_1+2r}\ e^{2a|s|^g}\ \|y_1 - y_2\|, \qquad s \in S(R).$$

Hence we deduce the inequality

$$\|\psi(s,y_1(s)) - \psi(s,y_2(s))\|^{S(R)}_{g,a,r-v} \leq K\ R^{r_1+r+v}\ e^{aR^g}\ \|y_1 - y_2\|, \tag{15.7}$$

valid for all $r \leq \min\{-r_0', -v-r_1\}$. Putting $y_1 = y$ and $y_2 = 0$ and noting that $\psi(s,0) = 0$, we see that

$$\psi(s,y(s)) \in B_{g,a,r-v}(S(R))$$

whenever $y \in E$. In view of (15.5) and the conclusions of proposition 4.3 we can define a mapping T from E into $B_{g,a,r}(S(R))$ by

$$Ty(s) = \Lambda^{S(R)}_{g,a,r-v}\ [\psi(s,y(s)) + \varphi_0(s)],$$

provided that $r \leq \min\{-r_0', -v-r_1\}$, $|a|$ is a sufficiently small and R a sufficiently large positive number. From (15.7) and the properties of $\Lambda^{S(R)}_{g,a,r-v}$ it follows, moreover, that T is a contraction on E if R is larger than some positive number R_0. Then there is a unique element y of E with the property that

$$y(s) = \Lambda^{S(R)}_{g,a,r-v}\ [\psi(s,y(s)) + \varphi_0(s)] \qquad (R \geq R_0). \tag{15.8}$$

It is easily verified that the constant R_0 can be chosen independent of r. We shall now show that $y \in B_{g,a,r}(S(R))$ for all $r \leq 0$. Let $r < \min\{-r_0', -v-r_1\}$ and suppose that $y \in B_{g,a,r}(S(R))$. Then it follows from (15.6) that

$$\|\psi(s,y(s))\|^{S(R)}_{g,a,r'-v} \leq K\ \|y\|^2, \qquad R \geq R_0,$$

where

$$r' = 2r + r_1 + v < r.$$

Utilizing the 3rd property mentioned in proposition 4.3 we find

$$\Lambda^{S(R)}_{g,a,r-v}[\psi(s,y(s))+\varphi_o(s)]\in B_{g,a,r'}(S(R)), \qquad R\geq R_o,$$

and consequently,

$$y\in B_{g,a,r'}(S(R)), \qquad R\geq R_o.$$

By means of induction we conclude that $y\in B_{g,a,r}(S(R))$ for all $r<\min\{-r_o', -v-r_1\}$ and, hence, for $r\leq 0$.

Obviously, the function y is a solution of (15.3), and therefore of (15.2), in S(R). Now, let S' be a closed subsector of S, $S'=S[\alpha_1',\alpha_2']$. Since S(R) is an S-proper region there exists a number $R'\geq R$ such that $S_{R'}(\alpha_1',\alpha_2')\subset S(R)$. Hence it follows that $y\in B_{g,a,r}(S_{R'}(\alpha_1',\alpha_2'))$ for all $r\leq 0$ and thus $y\in A_{go}(S')$. Next suppose that S' is a closed subsector of S_o such that $S\subset S'\subset S_o$. According to the result stated directly after proposition 4.3 (with the roles of S and S' interchanged!) there exist a set of S'-proper regions $\{S'(R),\ R>0\}$ such that $S(R)\subset S'(R)$ for all $R>0$, and linear mappings $\Lambda^{S'(R)}_{g,a,r-v}$ possessing the following property in addition to those mentioned in proposition 4.3,

$$\Lambda' f|_{S(R)} = \Lambda(f|_{S(R)}), \qquad f\in B_{g,a,r-v}(S'(R)), \tag{15.9}$$

where we have put $\Lambda^{S(R)}_{g,a,r-v}=\Lambda$ and $\Lambda^{S'(R)}_{g,a,r-v}=\Lambda'$. Let E' denote the unit sphere of $B_{g,a,r}(S'(R))$. In the manner described above for S one obtains a function $y'\in E'$ satisfying the equation

$$y'=\Lambda'[\psi(s,y'(s))+\varphi_o(s)],$$

provided that $r\leq\min\{-r_o', -v-r_1\}$, $|a|$ is a sufficiently small and R a sufficiently large positive number. With (15.9) we find

$$y'|_{S(R)} = \Lambda[\psi(s,y'(s))+\varphi_o(s)]$$

Clearly, $y'|_{S(R)}\in E$. As (15.8) has a unique solution in E, we conclude that $y'|_{S(R)}=y$. By analytic continuation of y one thus obtains a solution of (15.2) belonging to $A_{go}(S_o)$.

Finally, suppose that (iv) is satisfied. Let $S=S[\alpha_1,\alpha_2]$ again be a closed subsector of S_o such that (15.4) holds. As we pointed out before, this implies that the conditions of proposition 4.12 are fulfilled. Let $\{S(R),\ R>0\}$ be a suitable set of S-proper regions. If y is a solution of (15.2) belonging to $(A_{go}(S_o))^n$, then there exist a negative number a and a positive number R such that

$$y|_{S(R)}\in B_{g,a,r}(S(R))$$

for all $r \in \mathbb{R}$. Moreover, if $r \in \mathbb{R}$ is fixed, we can always achieve that $\|y\|_{g,a,r}^{S(R)} \leq 1$ by choosing a sufficiently large number R. By proposition 4.12 the linear mapping $\Lambda_{g,a,r}^{S(R)}$ is a left inverse of Δ if $|a|$ is sufficiently small, R sufficiently large and $r \leq -r_o$. Hence, by virtue of property 3, y satisfies (15.8) and the last statement of theorem 15.1 follows immediately.

Note that the condition that $\varphi \in V_g(S_o,0)$ is not necessary in this case. It can be replaced by the following two conditions: $\varphi \in V_o(S_o,0)$ and $\varphi_o \in (A_{go}(S_o))^n$.

II. Next, we consider the general case of theorem 15.1, i.e. $\tilde{f} \in (\mathbb{C}[[s^{-1/p}]]_{pg})^n$ and $\tilde{f}(\infty) = y_o$.

a) Suppose first that either $g = 0$, or $g \neq 0$ and $\alpha_2^o - \alpha_1^o \leq \frac{\pi}{g}$. Then, by theorem 3.2 there exists a vector function $f \in (A_g(S_o))^n$ such that $\hat{f} = \tilde{f}$. The transformation

$$y = z + f$$

takes (15.2) into the equivalent equation

$$z(s-1) = \tilde{\varphi}(s,z(s)), \tag{15.10}$$

where

$$\tilde{\varphi}(s,z) = \varphi(s,z+f(s)) - f(s-1).$$

By lemma 14.3,

$$\tilde{\varphi} \in V_g(S_o,0).$$

Further, we have

$$D_2\tilde{\varphi}(s,0) = D_2\varphi(s,f(s))$$

and thus

$$D_2\hat{\tilde{\varphi}}(s,0) = \tilde{A}(s).$$

Apparently, the function $\tilde{\varphi}$ satisfies the conditions of theorem 15.1. Moreover, equation (15.10) admits 0 as a formal solution. It now follows from the proof given in I that also in this case the conclusions of theorem 15.1 are true.

b) Now suppose that $g \notin k(A) \cup \{0,1\}$ and $\alpha_2^o - \alpha_1^o > \frac{\pi}{g}$. Let $S_1 = S(\alpha_1^o, \alpha_1^o + \frac{\pi}{g})$, $S_2 = S(\alpha_2^o - \frac{\pi}{g}, \alpha_2^o)$ and $S = S_1 \cap S_2 = S(\alpha_2^o - \frac{\pi}{g}, \alpha_1^o + \frac{\pi}{g})$. For all $k \in k(A)$ such that $k > g$ and $\alpha_1^o + \frac{\pi}{g} - \frac{\pi}{k} < \alpha_2^o - \frac{\pi}{g}$ we have

$$[\alpha_1^o + \frac{\pi}{g} - \frac{\pi}{k}, \alpha_2^o - \frac{\pi}{g}] \subset (\alpha_1^o, \alpha_2^o - \frac{\pi}{k}).$$

Since, by assumption, $(\alpha_1^o, \alpha_2^o - \frac{\pi}{k}) \cap \Sigma_k^- (\tilde{A}) = \phi$, we find that, with respect to the sector S, condition (iv) of the theorem is fulfilled.
From the discussion in II a) we conclude that equation (15.2) possesses two solutions, $y_1 \in (A_g(S_1))^n$ and $y_2 \in (A_g(S_2))^n$, with the property that $\hat{y}_1 = \hat{y}_2 = \tilde{f}$, and, furthermore, a unique solution $y_{12} \in (A_g(S))^n$ such that $\hat{y}_{12} = \tilde{f}$. Hence it follows that

$$y_1|_S = y_2|_S = y_{12}.$$

Consequently, the functions y_1 and y_2 define a solution $y \in (A_g(S_o))^n$ of (15.2), and $\hat{y} = \tilde{f}$. Clearly, the vector function with these properties is unique.
c) Finally, suppose that $g > 0$, $g \in k(A) \cup \{1\}$ and $\alpha_2^o - \alpha_1^o > \frac{\pi}{g}$. Let α_1 and α_2 be two real numbers such that

$$\alpha_1^o < \alpha_1 < \alpha_2^o - \frac{\pi}{g} \quad \text{and} \quad \alpha_1 + \frac{\pi}{g} < \alpha_2 < \alpha_2^o.$$

Observe that, for all $k \geq g$, we have

$$[\alpha_1 + \frac{\pi}{g} - \frac{\pi}{k}, \alpha_2 - \frac{\pi}{g}] \subset (\alpha_1^o, \alpha_2^o - \frac{\pi}{k}).$$

Arguing in the same manner as above, one proves the existence of a unique vector function $y_S \in (A_g(S))^n$ with the properties that y_S is a solution of (15.2) and $\hat{y}_S = \tilde{f}$. In view of the uniqueness of this function and the fact that α_1 and α_2 may be chosen arbitrarily close to α_1^o and α_2^o, respectively, it follows that, by analytic continuation of y_S one finally obtains a solution $y \in (A_g(S_o))^n$ with the property that $\hat{y} = \tilde{f}$. This completes the proof of theorem 15.1. □

Proposition 4.9 provides us with sufficient conditions so that the asymptotic expansion of a solution of (15.2) in a left or right sector remains valid when this solution is continued analytically in the direction of the positive or negative real axis, respectively. We shall use the same notation as in theorem 15.1 . Further, we define

$$S_1 = S[\alpha_1^o, \alpha_2^o), \quad S_2 = S(\alpha_1^o, \alpha_2^o]. \tag{15.11}$$

<u>THEOREM</u> 15.12. *Let* $j \in \{1,2\}$ *and let* α_1^o, α_2^o *be real numbers such that*

$$\alpha_j^o = m\pi, \quad m \in \mathbb{Z}$$

and

$$\pi < \alpha_2^o - \alpha_1^o \leq 2\pi.$$

Let $\varphi \in (V_g(S_j,y_o))^n$, *where* $g \in [0,1]$, $y_o \in \mathbb{C}^n$ *and* S_j *is defined by* (15.11). *With the notation of theorem* 15.1 *assume that*

(i) *equation* (15.2) *has a formal solution* $\tilde{f} \in (\mathbb{C}[\![s^{-1/p}]\!]_{pg})^n$ *with the property that* $\tilde{f}(\infty) = y_o$,

(ii) $\tilde{A}^{-1} \in \mathrm{End}(n;\mathbb{C}[\![s^{-1/p}]\!])$,

(iii) $k(\tilde{A}) \subset [g,1]$, $0 \notin k(\tilde{A})$ *and for all* $k \in k(\tilde{A})$ *and all* $\mu \in \mu_k^{\pm}(\tilde{A})$ *the following condition is satisfied*

$$\mathrm{Re}(\mu e^{ik\alpha_j^o}) < 0$$

Then (15.2) *possesses a solution* $y \in (A_g(S_j))^n$ *with the property that* $\hat{y} = \tilde{f}$. *If, in addition, condition* (iv) *of theorem* 15.1 *is fulfilled, then the solution with these properties is unique.*

<u>PROOF</u>: The proof of this theorem is for the greater part similar to that of theorem 15.1. We shall indicate here the chief modifications required.

1. $\tilde{f} = 0$. The sectors $S[\alpha_1,\alpha_2]$ occurring in part I of the proof of theorem 15.1 must be replaced by subsectors of S_j of the form $S[\alpha_1^o,\alpha_2]$ in case $j = 1$, and $S[\alpha_1,\alpha_2^o]$ if $j = 2$. Instead of proposition 4.3, 4.9 is to be used.

2. $\tilde{f} \neq 0$, $g(\alpha_2^o - \alpha_1^o) < \pi$. Then there exists a vector function $f \in (A_g(S_j))^n$ such that $\hat{f} = \tilde{f}$. In the manner described in part II.a) of the proof of theorem 15.1, (15.2) is transformed into an equivalent equation which admits 0 as a formal solution.

3. $\tilde{f} \neq 0$, $g \notin \{0,1\}$, $\alpha_2^o - \alpha_1^o \geq \frac{\pi}{g}$. Observe that condition (iii) implies that $\Sigma_k^{\pm}(\tilde{A}) \subset \bigcup_{\ell \in \mathbb{Z}} (\alpha_j^o - \frac{\pi}{k} + \frac{2\ell\pi}{k}, \alpha_j^o + \frac{2\ell\pi}{k})$ for all $k \in k(\tilde{A})$ such that $k \geq g$, and, consequently,

$$[\alpha_1^o, \alpha_2^o - \tfrac{\pi}{k}] \cap \Sigma_k^{\pm}(\tilde{A}) = \phi$$

for all those values of k. By theorem 15.1 there exists a function $y_1 \in (A_g(S(\alpha_1^o,\alpha_2^o))^n$ with the properties that y_1 is a solution of (15.2) and $\hat{y}_1 = \tilde{f}$. Let us suppose that $j = 1$. The case that $j = 2$ can be treated analogously. We choose a positive number ε in such a way that

$$\varepsilon < \frac{\pi}{g} - \frac{\pi}{k} \tag{15.13}$$

for all $k \in k(\tilde{A})$ such that $k > g$ and, moreover,

$$[\alpha_1^o - \varepsilon, \alpha_1^o] \cap \Sigma_g^{\pm}(\tilde{A}) = \phi \qquad \text{if} \quad g \in k(\tilde{A}). \tag{15.14}$$

Let S and S_{12} denote the subsectors of S_1 defined by

$$S = S[\alpha_1^o, \alpha_1^o + \frac{\pi}{g} - \varepsilon)$$

and

$$S_{12} = S \cap S(\alpha_1^o, \alpha_2^o) = S(\alpha_1^o, \alpha_1^o + \frac{\pi}{g} - \varepsilon).$$

Since the aperture of both sectors is less than $\frac{\pi}{g}$ there exist functions $y_2 \in (A_g(S))^n$ and $y_{12} \in (A_g(S_{12}))^n$ with the same properties as y_1 (cf. part 2 of this proof and theorem 15.1). Moreover, it follows from (15.13) and (15.14) that with respect to S_{12} condition (iv) of theorem 15.1 is fulfilled. Thus the vector function y_{12} is unique. Hence we conclude that y_1 and y_2 define a function $y \in (A_g(S_1))^n$ with the properties that y is a solution of (15.2) and $\hat{y} = \tilde{f}$.

4. $\tilde{f} \neq 0$, $g = 1$. In this case the method used above does not work, as $S(\alpha_1^o, \alpha_1^o + \pi - \varepsilon)$ is neither a left nor a right sector. However, the statements of the theorem can be proved by means of Laplace transform techniques as described in [6] and [17].

The following result applies to sectors S_o contained in a lower or an upper half plane. We use the same notation as in the preceding two theorems.

<u>THEOREM</u> 15.15. *Let* $j \in \{1,2\}$ *and let* α_1^o *and* α_2^o *be real numbers such that*

$$\alpha_j^o = m\pi, \qquad m \in \mathbb{Z}$$

and

$$0 < \alpha_2^o - \alpha_1^o \leq \pi.$$

If m *is even we shall use the upper sign in* (15.2), *if* m *is odd the lower sign. Let* $\varphi \in (V_g(S_j, y_o))^n$, *where* $g \in [0,1], y_o \notin C^n$ *and* S_j *is defined by* (15.11). *In addition*

to hypotheses (i) *and* (ii) *of theorem* 15.12 *assume that*
(iii)' *for all* $k \in k(\tilde{A})$ *such that* $k \neq 0$ *and* $k \geq g$ *and all* $\mu \in \mu_k^{\pm}(\tilde{A})$ *the following condition is satisfied*

$$\mathrm{Re}(\mu e^{ik\alpha_j^o}) > 0.$$

Then (15.2) *possesses a unique solution* $y \in (A_g(S_j))^n$ *with the property that* $\hat{y} = \tilde{f}$.

The proof of this theorem is based on the use of proposition 4.16 and follows the same line of argument as in the case of theorem 15.1. The modifications required are obvious and are left to the reader. For the same reason we shall omit the proof of the next, and last, theorem of this section, which concerns the case that $(D_2\varphi(s,\tilde{f}))^{-1} \notin \mathrm{End}(n\,;\,\mathbb{C}[[s^{-1/p}]])$ and is derived from proposition 4.11.

<u>THEOREM</u> 15.16. *Let* $S_o = S(\alpha_1^o,\alpha_2^o)$, *with* $\alpha_1^o = \mp \frac{\pi}{2}$, $\alpha_2^o = \alpha_1^o + \pi$, *and let* $\mathcal{S}$ *be a set of* S_o*-proper regions as defined in* §12.1. *Let* $\varphi \in (V_g(\mathcal{S},y_o))^n$, *where* $y_o \in \mathbb{C}^n$ *and* $g \in [0,1]$. *Suppose there exists a vector function* $f \in (A_g(\mathcal{S}))^n$ *with the properties that* $\hat{f}$ *is a formal solution of* (15.2) *and* $f(\infty) = y_o$. *Set*

$$D_2\hat{\varphi}(s,\hat{f}) = \tilde{A}$$

and assume that

(i) $\tilde{A} \in \hat{M}_o$
(ii) *for all* $k \in k(\tilde{A})$ *such that* $k \neq 0$ *and* $k \geq g$ *the following condition is satisfied*

$$[\alpha_2^o - \frac{\pi}{k}, \alpha_1^o] \cap \Sigma_k^{\pm}(\tilde{A}) = \phi.$$

Then equation (15.2) *possesses a unique solution* $y \in (A_g(\mathcal{S}))^n$ *with the property that* $\hat{y} = \hat{f}$.

<u>REMARK</u> 1. If $g < 1$ then the existence of a formal solution $\hat{f}$ of (15.2) such that $\hat{f}(\infty) = y_o$, according to theorem 3.2, is sufficient to guarantee the existence of a solution $f \in (A_g(\mathcal{S}))^n$ with the properties mentioned above. In the case that $g = 1$ such a result is not known, however (theorem 3.2 cannot be applied due to the fact that $\alpha_2^o - \alpha_1^o = \pi$). In that case the condition mentioned in theorem 15.16 is, a priori, stronger than the requirement that a formal solution of the equation exist.

<u>REMARK</u> 2. Since $\mathcal{S}$ is a set of S_o-proper regions, it follows that the solution found in theorem 15.16 is an element of $(A_g(S_o))^n$. It should be noted, however,

that the uniqueness of this solution in $(A_g(\mathbb{S}))^n$ does not imply uniqueness in $(A_g(S_o))^n$.

The results obtained in this section can be easily generalized to equations of the following form

$$\Phi(s,y(s),y(s\pm 1))=0. \tag{15.17}$$

Here the function Φ is supposed to possess analogous properties as the functions φ considered so far, with the only difference that Φ is a function of a 2n-dimensional vector (y,z) instead of the n-dimensional vector y and is assumed analytic in a neighbourhood of a point in $\mathbb{C}^{2n}$ of the form (y_o,y_o). Evidently, the coefficients of the expansion of $\hat{\Phi}$ into powers of (y,z) may be taken to belong to $(\mathbb{C}[[s^{-1/p}]])^n$ instead of $(\mathbb{C}[[s^{-1/p}]][s^{1/p}])^n$. The obvious analogue of the set $V_g(F,y_o)$ defined on p. 78 will be denoted by $V_g^{(2)}(F,y_o)$.

Theorems 15.1 - 15.16 can be rewritten without difficulty for equations of the form (15.17), provided the sets $V_g(F,y_o)$ (where the definition of F varies according to the theorem concerned) are replaced by the corresponding sets $V_g^{(2)}(F,y_o)$. The only additional assumption to be made is the following. Let $\tilde{f}$ be a formal solution of (15.17) such that $\tilde{f}(\infty)=y_o$. Then we assume that

$$D_3\hat{\Phi}(s,\tilde{f}(s),\tilde{f}(s\pm 1))\in\hat{M}_g.$$

Furthermore, the definition of $\tilde{A}$ is to be changed into

$$\tilde{A}(s)=-\{D_3\hat{\Phi}(s,\tilde{f}(s),\tilde{f}(s\pm 1))\}^{-1}D_2\hat{\Phi}(s,\tilde{f}(s),\tilde{f}(s\pm 1)).$$

With these modifications the conclusions of theorems 15.1 - 15.16 remain valid.

As an application of the generalized form of the theorems discussed in this section, let us consider the following problem. Suppose we seek a solution of equation (13.1) in a right instead of a left sector, or of (13.2) in a left instead of a right sector. Let $\varphi\in(V_g(S,y_o))^n$ where $\pm S$ is a left sector, $g\in[0,1]$, $y_o\in\mathbb{C}^n$, and let f be an element of $(A_g(S))^n$ with the properties that $\hat{f}$ is a formal solution of (15.2) and $f(\infty)=y_o$. The substitution $y=z+f$ takes (15.2) into

$$z(s\pm 1)=\varphi(s,z+f(s))-f(s\pm 1),$$

which may be rewritten in the form

$$z(s\pm 1)=\varphi(s,f(s))+D_2\varphi(s,f(s))z(s)+\psi(s,z(s)).$$

If $D_2\hat{\varphi}(s,\hat{f}) \in \hat{M}_o$ this equation can be transformed into

$$z(s\mp 1) = -D_2\varphi(s\mp 1,f(s\mp 1))^{-1}\{\varphi(s\mp 1,f(s\mp 1)) - z(s) + \psi(s\mp 1,z(s\mp 1))\}.$$

The resulting equation is a special case of (15.17) and can be treated by the methods discussed above.

§16. *Existence of formal solutions belonging to a Gevrey class.*

The results presented in this section concern the existence of formal solutions belonging to the Gevrey class $(\mathbb{C}[[s^{-1/p}]]_{pg})^n$ for some $g \in (0,1]$, of equations of the type (15.17). They have been derived primarily with a view to later application to block diagonalization problems in §17.

The approach we use is due to Ramis and we refer to his paper ([31]) for more detailed explanations.

Let $\mathbb{N}' = \mathbb{N} \cup \{0\}$. An element $y = \sum_{h\in\frac{1}{p}\mathbb{N}'} y_h s^{-h}$ of $\mathbb{C}[[s^{-1/p}]]$ defines a function $[y]$ on $\frac{1}{p}\mathbb{N}'$ such that

$$[y](h) = y_h, \qquad h \in \tfrac{1}{p}\mathbb{N}'.$$

Obviously, the mapping $y \to [y]$ is a bijection.

DEFINITION. *Let* g *and* α *be positive numbers and let* $\rho \in \mathbb{R}$. *By* $(b_{g,\alpha,\rho})^n$ *we denote the Banach space of* n*-dimensional vector functions* $a : \frac{1}{p}\mathbb{N}' \to \mathbb{C}^n$ *with the property that*

$$\sup_{h\in\frac{1}{p}\mathbb{N}'} \Gamma(1+h)^{-1/g}\,\alpha^{-h}(1+h)^{-\rho}\,|a(h)| < \infty.$$

Observe that $\sum_{h\in\frac{1}{p}\mathbb{N}'} a(h)\, s^{-h} \in (\mathbb{C}[[s^{-1/p}]]_{pg})^n$ if and only if $a \in (b_{g,\alpha,\rho})^n$ for some $\alpha > 0$ and $\rho \in \mathbb{R}$.

The norm on $(b_{g,\alpha,\rho})^n$ will be denoted by $\|.\|_{g,\alpha,\rho}$. If no confusion is possible, we shall simply write $\|.\|$.

The following lemma shows that $b_{g,\alpha,\rho}$ is an algebra.

LEMMA 16.1. *Let* g *and* α *be positive numbers and* $\rho \in \mathbb{R}$. *There exists a positive constant* K *independent of* α *such that*

$$\|a\,b\|_{g,\alpha,\rho} \leq K\ \|a\|_{g,\alpha,\rho}\ \|b\|_{g,\alpha,\rho}$$

for all pairs $a,b \in b_{g,\alpha,\rho}$.

PROOF: Noting that

$$a\,b\,(h) = \sum_{\ell+k=h} a(\ell)\,b(k)$$

we find

$$|ab(h)| \leq \|a\|\ \|b\|\ \Gamma(1+h)^{1/g}\ \alpha^{h}(1+h)^{\rho} \sum_{\ell+k=h} \left\{\frac{\Gamma(1+\ell)\Gamma(1+k)}{\Gamma(1+h)}\right\}^{1/g} \left(1+\frac{\ell k}{1+h}\right)^{\rho}.$$

The finite sum on the right-hand side of this inequality closely resembles the sum occurring in the proof of lemma 14.1 and can be estimated in a similar manner. It turns out to be bounded by a constant independent of h. □

We are now going to study the action of some elementary operators on the Banach spaces $(b_{g,\alpha,\rho})^n$.

Let τ_+ and τ_- be the linear operators defined by

$$\tau_{\pm} f(s) = f(s \pm 1), \qquad f \in (\mathbb{C}[\![s^{-1/p}]\!])^n,$$

and let

$$\delta_{\pm} = \tau_{\pm} - 1.$$

The corresponding operators on $(b_{g,\alpha,\rho})^n$ will be denoted by the same symbols. Let $r \in \frac{1}{p}\mathbb{Z}$. By $[s^r]$ we shall denote the linear mapping defined by

$$[s^r][f] = [s^r f], \qquad f \in s^{-r}(\mathbb{C}[\![s^{-1/p}]\!])^n \cap (\mathbb{C}[\![s^{-1/p}]\!])^n.$$

Let $a \in b_{g,\alpha,\rho}$. We have

$$\tau_{\pm}(a)(h) = \sum_{\ell=0}^{[h-1/p]} (\mp 1)^{\ell} \binom{h-1}{\ell}\, a(h-\ell), \qquad h \in \frac{1}{p}\,\mathbb{N}'.$$

Hence we derive the inequality

$$\|\tau_\pm(a)\|_{g,\alpha,\rho} \le \|a\| \sup_{h\in\frac{1}{p}\mathbb{N}'} \sum_{\ell=0}^{[h-1/p]} \left\{\frac{\Gamma(1+h-\ell)}{\Gamma(1+h)}\right\}^{1/g-1} \frac{h-\ell}{h} \frac{\alpha^{-\ell}}{\ell!} \left(\frac{1+h-\ell}{1+h}\right)^{\rho}.$$

It follows that

$$\|\tau_\pm(a)\|_{g,\alpha,\rho} \le C_\tau(\alpha,\rho)\ \|a\|, \tag{16.2}$$

where

$$C_\tau(\alpha,\rho) = \sum_{\ell=0}^{\infty} (1+\ell)^{|\rho|} \frac{\alpha^{-\ell}}{\ell!}. \tag{16.3}$$

Let $r\in\frac{1}{p}\mathbb{Z}$. If $r>0$, suppose that $a(h)=0$ for all $h<r$. Then we have

$$\begin{aligned}[s^r]a(h) &= 0 && \text{for all } h<-r \quad \text{if } r\le 0,\\ &= a(h+r) && \text{for all } h\ge \max\{0,-r\}.\end{aligned}$$

One easily verifies that

$$\|[s^r]\,a\|_{g,\alpha,\rho+r/g} \le C(r,g,\rho)\ \alpha^r\ \|a\|, \tag{16.4}$$

where $C(r,g,\rho)$ is a positive number independent of α.
Now let $r\in[0,1]\cap\frac{1}{p}\mathbb{Z}$. A simple calculation shows that

$$\begin{aligned}[s^r]\delta_\pm(a)(h) &= 0 && \text{for all } h<1+\frac{1}{p}-r,\\ &= \sum_{\ell=1}^{[h+r-1/p]} (\mp 1)^\ell \binom{h+r-1}{\ell}\, a(h+r-\ell) && \text{for all } h\ge 1+\frac{1}{p}-r.\end{aligned} \tag{16.5}$$

Hence one obtains the following estimate

$$\|[s^r]\delta_\pm(a)\|_{g,\alpha,\rho+1-(1-r)/g} \le C_{\delta,r}(g,\alpha,\rho)\|a\| \tag{16.6}$$

where

$$C_{\delta,r}(g,\alpha,\rho) = C_g\ \alpha^{r-1} \sum_{\ell=1}^{\infty} (1+\ell)^{|\rho+1|} \frac{\alpha^{-(\ell-1)}}{\ell!}. \tag{16.7}$$

C_g is a positive number independent of α.

LEMMA 16.8. *Let* $g \in (0,1] \cap \frac{1}{p}\mathbb{Z}$, $\rho \in \mathbb{R}$, $A_o \in G\ell(n;\mathbb{C})$, $B_o \in \mathrm{End}(n;\mathbb{C})$ *and let*

$$\Delta_\pm = [s^{1-g}]B_o\delta_\pm + A_o.$$

There exists a positive number $\alpha(A_o,B_o)$ *such that for all* $\alpha > \alpha(A_o,B_o)$ *the mapping*

$$(b_{g,\alpha,\rho})^n \xrightarrow{\Delta_\pm} (b_{g,\alpha,\rho})^n$$

is a bijection.

PROOF: Let $a \in (b_{g,\alpha,\rho})^n$ and let $\Delta_\pm a = b$. From (16.6) it follows that $b \in (b_{g,\alpha,\rho})^n$. Using (16.5) we obtain a recursive relation for a(h),

$$\begin{aligned} a(h) &= A_o^{-1}b(h) && \text{if } h < g + \frac{1}{p}, \\ &= A_o^{-1}\{b(h) - B_o \sum_{\ell=1}^{[h+1-g-1/p]} (\mp 1)^\ell \binom{h-g}{\ell} a(h+1-g-\ell)\} && \text{if } h \geq g + \frac{1}{p}. \end{aligned} \tag{16.9}$$

Obviously, $b = 0$ implies $a = 0$ and thus $\Delta_\pm$ is injective. We shall prove the surjectivity by means of induction on h, as follows. Suppose that $b \in (b_{g,\alpha,\rho})^n$. Let $h > g + \frac{1}{p}$. Assume that there is a positive number $C > |A_o^{-1}|\ \|b\|$ such that $|a(k)| \leq C\,\Gamma(1+k)^{1/g}\alpha^k(1+k)^\rho$ for all $k < h$. With (16.9) and (16.6) we then have

$$|a(h)|\Gamma(1+h)^{-1/g}\alpha^{-h}(1+h)^{-\rho} \leq |A_o^{-1}|\{\|b\| + |B_o|c_{\delta,1-g}(g,\alpha,\rho)C\}.$$

It is apparent from (16.7) that $\lim_{\alpha\to\infty} c_{\delta,1-g}(g,\alpha,\rho) = 0$. Consequently, there exists a positive number $\alpha(A_o,B_o)$ such that

$$|B_o|c_{\delta,1-g}(g,\alpha,\rho) < |A_o^{-1}|^{-1}$$

for all $\alpha > \alpha(A_o,B_o)$. If C is chosen in such a way that

$$C \geq \frac{|A_o^{-1}|\,\|b\|}{1 - |A_o^{-1}|\,|B_o|c_{\delta,1-g}(g,\alpha,\rho)}$$

then it follows that

$$|a(h)|\Gamma(1+h)^{-1/g}\alpha^{-h}(1+h)^{-\rho} \leq C.$$

Hence we conclude, by means of induction, that

$$a \in (b_{g,\alpha,\rho})^n,$$

which shows that $\Delta_\pm$ is a surjection as well. □

Next we consider 'small' perturbations of several of the operators introduced above. To this end we use the concept of Fredholm operator. A continuous linear mapping L from the Banach space X into the Banach space Y is called a Fredholm operator if ker L and coker L have finite dimension. The difference of these two dimensions is the index of L, denoted by $\chi(L;X,Y)$:

$$\chi(L;X,Y) = \dim \ker L - \dim \operatorname{coker} L.$$

(If no confusion is possible we shall write $\chi(L)$).
The composition of two Fredholm operators $L_1 : X \longrightarrow Y$ and $L_2 : Y \longrightarrow Z$ is again a Fredholm operator and we have

$$\chi(L_2 \circ L_1;\ X,Z) = \chi(L_1;\ X,Y) + \chi(L_2;\ Y,Z).$$

Without proofs we mention two important stability theorems (cf.[33]).

<u>THEOREM</u> 16.10. *Let* X *and* Y *be Banach spaces and let* $L \in B(X,Y)$ [8)] *be a Fredholm operator. There exists a positive number* λ *with the following property: if* $K \in B(X,Y)$ *and* $\|K\| < \lambda$ *then* $L + K$ *is a Fredholm operator and*

$$\chi(L + K) = \chi(L).$$

<u>THEOREM</u> 16.11 . *Let* X *and* Y *be Banach spaces and let* $L \in B(X,Y)$ *be a Fredholm operator. If* $K \in B(X,Y)$ *is compact, then* $L + K$ *is a Fredholm operator and*

$$\chi(L + K) = \chi(L).$$

Now let $g > 0$, $\alpha > 0$, ρ, ρ_1 and $\rho_2 \in \mathbb{R}$. It is not difficult to verify that the natural injection

$$(b_{g,\alpha,\rho_2})^n \longrightarrow (b_{g,\alpha,\rho_1})^n$$

is compact if $\rho_2 < \rho_1$. Application of theorem 16.11 leads to the following result.

<u>LEMMA</u> 16.12. (cf. [31], proposition 1.3.7). *Let* $g > 0$, $\alpha > 0$, ρ, ρ_1 *and* $\rho_2 \in \mathbb{R}$. *Suppose that* $\rho_2 < \rho_1$. *If* $L \in B((b_{g,\alpha,\rho})^n, (b_{g,\alpha,\rho_1})^n)$ *is a Fredholm operator and* $K \in B((b_{g,\alpha,\rho})^n, (b_{g,\alpha,\rho_2})^n)$, *then* $L + K$ *is a Fredholm operator and*

$$\chi(L + K; (b_{g,\alpha,\rho})^n, (b_{g,\alpha,\rho_1})^n) = \chi(L; (b_{g,\alpha,\rho})^n, (b_{g,\alpha,\rho_1})^n).$$

---

8) B(X,Y) denotes the set of continuous linear operators from X into Y.

LEMMA 16.13. *Let* $g \in (0,1]$, $\alpha > 0$, $\rho \in \mathbb{R}$ *and let* $A, B \in \mathrm{End}(n; b_{g,\alpha,\rho-1/pg})$. *Let*

$$\Delta_{\pm} = B\tau_{\pm} + A. \tag{16.14}$$

Assume that either of the two matrices A(0) *and* B(0) *is the identity while the other one is nilpotent. Then the mapping*

$$(b_{g,\alpha,\rho})^n \xrightarrow{\Delta_{\pm}} (b_{g,\alpha,\rho})^n$$

is a bijection.

PROOF: Let $A(0) = A_o$, $B(0) = B_o$, $[s^{1/p}](A - A_o) = \widetilde{A}$ and $[s^{1/p}](B - B_o) = \widetilde{B}$. From (16.14) and lemma 16.1 it follows that $\widetilde{A}$ and $\widetilde{B} \in \mathrm{End}(n; b_{g,\alpha,\rho})$. Let $a \in (b_{g,\alpha,\rho})^n$. Putting $\widetilde{B}\tau_{\pm} + \widetilde{A} = \widetilde{\Delta}_{\pm}$ we have

$$\Delta_{\pm} a(h) = (B_o + A_o)a(h) + B_o \delta_{\pm} a(h) + [s^{-1/p}]\widetilde{\Delta}_{\pm} a(h), \qquad h \in \frac{1}{p}\mathbb{N}'.$$

The assumptions made in the proposition imply that $B_o + A_o$ is invertible. Hence the injectivity of the mapping $\Delta_{\pm}$ is apparent. Suppose that A_o is nilpotent and $B_o = I$. The other case can be treated analogously. Obviously,

$$\chi(B_o\tau_{\pm}, (b_{g,\alpha,\rho})^n, (b_{g,\alpha,\rho})^n) = 0.$$

Now the norm of A_o can be made as small as we like by means of a transformation of the form $A_o \to C^{-1}A_oC$, where $C \in G\ell(n;\mathbb{C})$. Such a transformation, when applied to $B_o\tau_{\pm} + A_o$, clearly does not change the index of this operator. With the aid of theorem 16.10 we conclude that

$$\chi(B_o\tau_{\pm} + A_o, (b_{g,\alpha,\rho})^n, (b_{g,\alpha,\rho})^n) = 0.$$

Furthermore, it is easily seen that the mapping

$$(b_{g,\alpha,\rho})^n \xrightarrow{[s^{-1/p}]\widetilde{\Delta}_{\pm}} (b_{g,\alpha,\rho-1/pg})^n$$

is bounded and thus, by lemma 16.12 ,

$$\chi(\Delta_{\pm}, (b_{g,\alpha,\rho})^n, (b_{g,\alpha,\rho})^n) = 0.$$

COROLLARY TO LEMMA 16.13. *Let* $g \in (0,1]$, $\alpha > 0$, $\rho \in \mathbb{R}$. *Let* $A \in \mathrm{End}(n; b_{g,\alpha,\rho})$. *If* A(0) *is invertible then*

$$A^{-1} \in \mathrm{End}(n; b_{g,\alpha,\rho+1/pg}).$$

PROPOSITION 16.15. *Let* $k \in (0,1] \cap \frac{1}{p}\mathbb{Z}$, $g \in (0,k]$, $\alpha > 0$, $\rho \in \mathbb{R}$ *and* $A, B \in End(n; b_{g,\alpha,\rho-1/pg})$. *Let*

$$\Delta_{\pm} = [s^{1-k}]B\delta_{\pm} + A \tag{16.16}$$

Assume that

(i) $A(0) \in G\ell(n;\mathbb{C})$

(ii) *if* $g = k$ *then* $\alpha > \alpha(A(0),B(0))$, *where* $\alpha(A(0),B(0))$ *is the positive number mentioned in lemma* 16.8 .

Then the mapping

$$(b_{g,\alpha,\rho})^n \xrightarrow{\Delta_{\pm}} (b_{g,\alpha,\rho})^n$$

is a bijection.

PROOF: The injectivity of the mapping is easily verified with the aid of formula (16.5). From (16.6) we conclude that the mapping

$$(b_{g,\alpha,\rho})^n \xrightarrow{[s^{1-k}]B\delta_{\pm}} (b_{g,\alpha,\rho+1-k/g})^n$$

is bounded. Hence, if $g < k$ then by lemma 16.12,

$$\chi(\Delta_{\pm}, (b_{g,\alpha,\rho})^n, (b_{g,\alpha,\rho})^n) = \chi(A, (b_{g,\alpha,\rho})^n, (b_{g,\alpha,\rho})^n) = 0.$$

Now suppose that $g = k$. In view of lemma 16.8, the assumptions made in the proposition imply that

$$\chi([s^{1-k}]B(0)\delta_{\pm} + A(0); (b_{k,\alpha,\rho})^n, (b_{k,\alpha,\rho})^n) = 0.$$

Let $[s^{1/p}](A - A(0)) = \widetilde{A}$. Then $\widetilde{A} \in End(n; b_{k,\alpha,\rho})$. It follows from (16.4) that the mapping

$$(b_{k,\alpha,\rho})^n \xrightarrow{[s^{-1/p}]\widetilde{A}} (b_{k,\alpha,\rho-1/pk})^n$$

is bounded. Application of lemma 16.12 now yields

$$\chi(\Delta_{\pm}; (b_{k,\alpha,\rho})^n, (b_{k,\alpha,\rho})^n) = 0. \qquad \square$$

LEMMA 16.17. *Let* $g \in (0,1]$, $\rho \in \mathbb{R}$. *There exists a nonnegative number* α_o, *equal to* 0 *if* $g < 1$, *such that, for all* $\alpha > \alpha_o$, *the linear mapping*

$$(b_{g,\alpha,\rho})^n \xrightarrow{[s]\delta_\pm} (b_{g,\alpha,\rho+1})^n$$

is a Fredholm operator, and

$$\chi([s]\delta_\pm;(b_{g,\alpha,\rho})^n,(b_{g,\alpha,\rho+1})^n)=0.$$

PROOF: It is apparent from (16.6) that $[s]\delta_\pm \in B((b_{g,\alpha,\rho})^n,(b_{g,\alpha,\rho+1})^n)$. Let L_1 be the linear mapping defined by: $L_1 a(h) = \mp h\, a(h)$, $h \in \frac{1}{p}\mathbb{N}$, and let $L_2 = [s]\delta_\pm - L_1$. Obviously, $L_1 \in B((b_{g,\alpha,\rho})^n,(b_{g,\alpha,\rho+1})^n)$ and dim ker $(L_1) =$ dim coker $(L_1) = n$. With the aid of (16.5) one readily verifies that $L_2 \in B((b_{g,\alpha,\rho})^n,(b_{g,\alpha,\rho+2-1/g})^n)$. Applying lemma 16.12 one finds that $\chi([s]\delta_\pm;(b_{g,\alpha,\rho})^n,(b_{g,\alpha,\rho+1})^n) = \chi(L_1) = 0$ if $g<1$.

Now suppose that $g=1$. Let $b \in (b_{1,\alpha,\rho+1})^n$ and consider the equation

$$[s]\delta_\pm y = b. \tag{16.18}$$

Using (16.5) we obtain the relations

$$b(0)=0$$

and

$$b(h) = \sum_{\ell=1}^{[h+1-1/p]} (\mp 1)^\ell \binom{h}{\ell}\, y(h+1-\ell), \qquad h \in \frac{1}{p}\mathbb{N}.$$

Hence we derive a recursive relation for $y(h)$:

$$y(h) = \mp\frac{1}{h}\{b(h) - \sum_{\ell=2}^{[h+1-1/p]} (\mp 1)^\ell \binom{h}{\ell}\, y(h+1-\ell)\} = \mp\frac{1}{h}\{b(h) - L_2 y(h)\}, \qquad h \in \frac{1}{p}\mathbb{N}.$$

It follows that $[s]\delta_\pm y = 0$ if and only if $y(h)=0$ for all $h \in \frac{1}{p}\mathbb{N}$, and thus dim ker $([s]\delta_\pm;(b_{1,\alpha,\rho})^n,(b_{1,\alpha,\rho+1})^n) = n$. A straightforward calculation yields the inequality

$$\|L_2 y\|_{1,\alpha,\rho+1} \le C_1(\alpha,\rho)\ \|y\|_{1,\alpha,\rho}\ , \tag{16.19}$$

where

$$C_1(\alpha,\rho) = \alpha^{-1} \sum_{\ell=2}^{\infty} \ell^{|\rho+1|}\ \frac{\alpha^{2-\ell}}{\ell!}\ . \tag{16.20}$$

Since $\lim_{\alpha\to\infty} C_1(\alpha,\rho) = 0$, there exists a positive number α_0 such that

$C_1(\alpha,\rho) < (p+1)^{-1}$ for all $\alpha > \alpha_o$. By the same reasoning we used in the proof of lemma 16.8 it can be shown that $y \in (b_{1,\alpha,\rho})^n$ if $\alpha > \alpha_o$. Thus, equation (16.18) has a solution in $(b_{1,\alpha,\rho})^n$ provided $b(0) = 0$. Then

$\dim \operatorname{coker}([s]\delta_\pm; (b_{1,\alpha,\rho})^n, (b_{1,\alpha,\rho+1})^n) = n$ and, consequently,

$\chi([s]\delta_\pm; (b_{1,\alpha,\rho})^n, (b_{1,\alpha,\rho+1})^n) = 0.$ □

PROPOSITION 16.21. *Let* $g \in (0,1]$, $\alpha > 0$, $\rho \in \mathbb{R}$ *and* $A \in \operatorname{End}(n; b_{g,\alpha,\rho})$. *Let*

$$\Delta_\pm = [s]\delta_\pm + A.$$

Assume that

(i) $A(0) \mp h I \in G\ell(n;\mathbb{C})$ *for all* $h \in \frac{1}{p}\mathbb{N}$,

(ii) $\alpha > \alpha_o$, *where* α_o *is the nonnegative number mentioned in lemma* 16.17.

If, in addition, $A(0) \in G\ell(n;\mathbb{C})$, *then the mapping*

$$(b_{g,\alpha,\rho})^n \xrightarrow{\Delta_\pm} (b_{g,\alpha,\rho+1})^n$$

is a bijection. If, on the other hand, $A(0) \notin G\ell(n;\mathbb{C})$ *and if* y_o *is an eigenvector of* $A(0)$ *corresponding to the eigenvalue* 0, *then there is a* $y \in \ker(\Delta_\pm; (b_{g,\alpha,\rho})^n, (b_{g,\alpha,\rho+1})^n)$ *such that* $y(0) = y_o$.

PROOF: By lemma 16.12, $\chi(\Delta_\pm; (b_{g,\alpha,\rho})^n, (b_{g,\alpha,\rho+1})^n) = \chi([s]\delta_\pm; (b_{g,\alpha,\rho})^n, (b_{g,\alpha,\rho+1})^n)$. Hence, in view of the preceding lemma,

$$\chi(\Delta_\pm; (b_{g,\alpha,\rho})^n, (b_{g,\alpha,\rho+1})^n) = 0 \tag{16.22}$$

if $\alpha > \alpha_o$. Now let $b \in (b_{g,\alpha,\rho+1})^n$ and consider the equation $\Delta_\pm\, y = b$. This is equivalent to the set of relations

$$A(0)y(0) = b(0),$$

$$(A(0) \mp h I)y(h) + \sum_{\ell=2}^{[h+1-1/p]} (\mp 1)^\ell \binom{h}{\ell} y(h+1-\ell) + \sum_{k=1/p}^{h} A(k)y(h-k) = b(h),$$

$$h \in \frac{1}{p}\mathbb{N}.$$

If $A(0) \mp h I \in G\ell(n;\mathbb{C})$ for all $h \in \frac{1}{p}\mathbb{N}'$, then, obviously,

$\ker(\Delta_\pm;(b_{g,\alpha,\rho})^n,(b_{g,\alpha,\rho+1})^n) = \{0\}$, and thus $\Delta_\pm$ is a bijection from $(b_{g,\alpha,\rho})^n$ onto $(b_{g,\alpha,\rho+1})^n$, provided $\alpha > \alpha_o$. If, on the other hand, $A(0) \mp hI \in G\ell(n;\mathbb{C})$ for all $h \in \frac{1}{p}\mathbb{N}$, but $A(0)$ is not invertible, then, as the above relations show, y is determined uniquely by b and $y(0)$. Note that $y \in \ker(\Delta_\pm;(b_{g,\alpha,\rho})^n,(b_{g,\alpha,\rho+1})^n)$ implies $y(0) \in \ker(A(0))$. Hence we conclude that
$\dim\ker(\Delta_\pm;(b_{g,\alpha,\rho})^n,(b_{g,\alpha,\rho+1})^n) \leq \dim\ker(A(0))$. Further, it is obvious that $\dim\operatorname{coker}(\Delta_\pm;(b_{g,\alpha,\rho})^n,(b_{g,\alpha,\rho+1})^n) \geq \dim\operatorname{coker}(A(0)) = \dim\ker(A(0))$. On account of (16.22) the dimensions of $\ker(\Delta_\pm)$ and $\operatorname{coker}(\Delta_\pm)$ must be equal, hence

$$\dim\ker(\Delta_\pm;(b_{g,\alpha,\rho})^n,(b_{g,\alpha,\rho+1})^n) = \dim\ker(A(0)).$$

The second statement of the proposition now follows immediately. □

We now proceed to study nonlinear equations with coefficients in $(b_{g,\alpha,\rho})^n$.

<u>DEFINITION</u>. *Let* $g>0$, $\alpha>0$ *and* $\rho\in\mathbb{R}$. *By* $\hat{V}^{(2)}_{g,\alpha,\rho}(y_o)$ *we shall denote the set of functions* φ *that can be represented by a series of the form*

$$\varphi(y,z) = \sum_{\substack{|m|\geq 0\\ |\ell|\geq 0}} \varphi_{m\ell}\,(y-y_o)^m(z-y_o)^\ell, \tag{16.23}$$

where $y_o\in\mathbb{C}^n$, $m=(m_1,\ldots,m_n)$, $\ell=(\ell_1,\ldots,\ell_n)$ *and the coefficients* $\varphi_{m\ell}$ *are elements of* $b_{g,\alpha,\rho}$ *with the following property: there exist positive numbers* δ_o *and* C *such that*

$$\|\varphi_{m\ell}\|_{g,\alpha,\rho} \leq C\,\delta_o^{-|m+\ell|} \tag{16.24}$$

for all m,ℓ *such that* $|m|\geq 0$, $|\ell|\geq 0$.

We are going to consider the following equation

$$\varphi(y,\tau_\pm y) = 0.$$

Before stating the main result we give two simple lemmas.

LEMMA 16.25. *Let* $\varphi \in (\hat{V}^{(2)}_{g,\alpha,\rho}(y_0))^n$, *where* $g>0$, $\alpha>0$, $\rho \in \mathbb{R}$ *and* $y_0 \in \mathbb{C}^n$. *If* $y \in (b_{g,\alpha,\rho})^n$ *and* $\|y-y_0\|_{g,\alpha,\rho} \leq \delta_1$, *where* δ_1 *is a sufficiently small positive number, then*

$$\varphi(y,\tau_\pm y) \in (b_{g,\alpha,\rho})^n.$$

PROOF: From lemma 16.1 we deduce that

$$\|\varphi_{m\ell}(y-y_0)^m(\tau_\pm y-y_0)^\ell\|_{g,\alpha,\rho} \leq K^{|m+\ell|}\|\varphi_{m\ell}\| \; \|y-y_0\|^{|m|}\|\tau_\pm y-y_0\|^{|\ell|}.$$

Noting that $\tau_\pm y-y_0=\tau_\pm(y-y_0)$ and using (16.2) and (16.24) we obtain

$$\|\sum_{\substack{|m|\geq 0\\|\ell|\geq 0}}\varphi_{m\ell}(y-y_0)^m(\tau_\pm y-y_0)^\ell\|_{g,\alpha,\rho} \leq \sum_{\substack{|m|\geq 0\\|\ell|\geq 0}} C\left(\frac{K}{\delta_0}\|y-y_0\|\right)^{|m+\ell|}C_\tau(\alpha,\rho)^{|\ell|}.$$

It follows that $\|\varphi(y,\tau_\pm y)\|_{g,\alpha,\rho}<\infty$ if $\|y-y_0\|<\dfrac{\delta_0}{K\,C_\tau(\alpha,\rho)}$.

LEMMA 16.26. *Let*

$$\psi(y,z)=\sum_{|m+\ell|\geq 2}\varphi_{m\ell}(y-y_0)^m(z-y_0)^\ell,$$

where $\varphi_{m\ell}\in(b_{g,\alpha,\rho})^n$ *and* (16.24) *is supposed to hold for all* m,ℓ *such that* $|m+\ell|\geq 2$. *There exists a positive number* $\delta_2\leq\delta_1$ *such that, for all* y_1, $y_2\in(b_{g,\alpha,\rho})^n$ *with* $\|y_1-y_0\|_{g,\alpha,\rho}\leq\delta_2$, $\|y_2-y_0\|_{g,\alpha,\rho}\leq\delta_2$, *the following inequality holds*

$$\|\psi(y_1,\tau_\pm y_1)-\psi(y_2,\tau_\pm y_2)\|_{g,\alpha,\rho}\leq D\max\{\|y_1\|,\|y_2\|\}\;\|y_1-y_2\|,$$

where $D=CK\delta_0^{-2}\sum_{|m+\ell|\geq 2}|m+\ell|\;C_\tau(\alpha,\rho)^{|\ell|}\left(\frac{K\delta_2}{\delta_0}\right)^{|m+\ell|-2}$ (K *is the constant mentioned in lemma* 16.1).

PROOF: It suffices to prove the following statement: if $u,v\in(b_{g,\alpha,\rho})^n$, $\|u\|\leq\delta$, $\|v\|\leq\delta$, then

$$\|u^m-v^m\|\leq\|u-v\|\max\{\|u\|,\|v\|\}\;|m|\;K\,(K\delta)^{|m|-2} \tag{16.27}$$

for all m such that $|m| \geq 2$. This will be demonstrated by means of induction on n. If $n=1$ we have, for all $m \geq 1$,

$$u^m - v^m = (u-v)\sum_{k=0}^{m-1} u^k\, v^{m-k-1}.$$

Application of lemma 16.1 yields the inequality

$$\|u^m - v^m\| \leq \|u-v\|\sum_{k=0}^{m-1} (\max\{\|u\|,\|v\|\})^{m-1}\, K^{m-1},\ m \geq 1, \tag{16.28}$$

and hence (16.27) follows immediately. Now suppose (16.27) is true for all $n' < n$. Let $u=(u_1,..,u_n)$, $v=(v_1,..,v_n)$, $m=(m_1,..,m_n)$ and write

$$u^m - v^m = u_1^{m_1}(u_2^{m_2}..\,u_n^{m_n} - v_2^{m_2}..\,v_n^{m_n}) + (u_1^{m_1} - v_1^{m_1})\, v_2^{m_2}..\,v_n^{m_n}. \tag{16.29}$$

Without loss of generality we may assume that $1 \leq m_1 < |m|$. Using (16.28) and lemma 16.1 we find for the second term on the right-hand side of (16.29),

$$\|(u_1^{m_1} - v_1^{m_1})\, v_2^{m_2}..\,v_n^{m_n}\| \leq \|u-v\|(\max\{\|u\|,\|v\|\})^{|m|-1}\, m_1\, K^{|m|-1}. \tag{16.30}$$

Suppose that $|m| - m_1 \geq 2$. Then, by assumption,

$$\|u_2^{m_2}..\,u_n^{m_n} - v_2^{m_2}..\,v_n^{m_n}\| \leq \|u-v\|\max\{\|u\|,\|v\|\}(|m|-m_1)K(K\delta)^{|m|-m_1-2}.$$

Hence we deduce that

$$\|u_1^{m_1}(u_2^{m_2}..\,u_n^{m_n} - v_2^{m_2}..\,v_n^{m_n})\| \leq \|u-v\|\max\{\|u\|,\|v\|\}(|m|-m_1)K(K\delta)^{|m|-2} \tag{16.31}$$

If, on the other hand, $|m| - m_1 = 1$, (16.27) follows directly by application of lemma 16.1. Combining (16.29),(16.30) and (16.31) we conclude that (16.27) is true. □

The main results of this section will be formulated in terms of formal power series in $s^{-1/p}$ instead of functions on the set $\frac{1}{p}\mathbb{N}'$. We recall that, corresponding to the series $y = \sum_{h\in\frac{1}{p}\mathbb{N}'} y_h\, s^{-h}$ we have a function $[y]$ on $\frac{1}{p}\mathbb{N}'$, defined by $[y](h) = y_h$ for all $h \in \frac{1}{p}\mathbb{N}'$, and vice versa. The same notation is used for vector and matrix functions.

DEFINITION. *Let* $y_o \in ¢^n$. *By* $\hat{V}_o^{(2)}(y_o)$ *we shall denote the set of all functions* φ *that can be represented by a series of the form* (16.23), *where the coefficients* $\varphi_{m\ell}$ *are elements of* $¢[\![s^{-1/p}]\!]$ *with the following property: there exists a positive constant* δ_o *such that*

$$\sup_{\{(m,\ell):|m|\geq 0,|\ell|\geq 0\}} \delta_o^{|m+\ell|}[\varphi_{m\ell}](h) < \infty \qquad \textit{for all } h \in \frac{1}{p}\mathbb{N}'.$$

If $\varphi \in (\hat{V}_o^{(2)}(y_o))^n$ *then* $[\varphi]$ *is defined by*

$$[\varphi](y,z) = \sum_{\substack{|m|\geq 0\\ |\ell|\geq 0}} [\varphi_{m\ell}](y-y_o)^m(z-y_o)^\ell.$$

Let $g>0$. *By* $\hat{V}_g^{(2)}(y_o)$ *we shall denote the set of all* $\varphi \in \hat{V}_o^{(2)}(y_o)$ *with the property that there exist a positive number* α *and a real number* ρ *such that*

$$[\varphi] \in \hat{V}_{g,\alpha,\rho}^{(2)}(y_o).$$

REMARK. Suppose that $\varphi \in V_g^{(2)}(S,y_o)$, where S is a sector, $y_o \in ¢^n$ and $g>0$. By means of computations similar to those used in deriving formula (14.5), it can be shown that $\hat{\varphi} \in \hat{V}_g^{(2)}(y_o)$.

THEOREM 16.32. *Let* $k \in (0,1] \cap \frac{1}{p}\mathbb{Z}$ *and let* $\varphi \in (\hat{V}_g^{(2)}(y_o))^n$, *where* $g \in [0,k]$ *and* $y_o \in ¢^n$. *Set*

$$D_1\varphi(y_o,y_o) + D_2\varphi(y_o,y_o) = A$$

and assume that

(i) $A \in s^{k-1}\, G\ell(n;¢[\![s^{-1/p}]\!])$

(ii) $\varphi_{m\ell} \in s^{k-1-1/p}(¢[\![s^{-1/p}]\!])^n$ *for all* m,ℓ *such that* $|m+\ell| \neq 1$.

Then the equation

$$\varphi(y,\tau_{\pm}y) = 0 \tag{16.33}$$

possesses a unique solution $y \in (¢[\![s^{-1/p}]\!]_{pg})^n$. *Moreover,* $y(\infty) = y_o$.

PROOF: The existence and uniqueness of a solution $y \in (¢[[s^{-1/p}]])^n$ of (16.33) can be proved by direct verification. In particular, it turns out that $y(\infty) = y_o$.

Now suppose that $g > 0$. There exist a positive number α and a real number ρ such that $[\varphi] \in (\hat{V}^{(2)}_{g,\alpha,\rho}(y_o))^n$. We put

$$[s^{1-k}A] = \tilde{A},\ [s^{1-k}\varphi_{oo}] = \tilde{\varphi}_{oo},\ [s^{1-k}\psi] = \tilde{\psi},$$

where ψ is defined as follows

$$\psi(y,z) = \sum_{|m+\ell| \geq 2} \varphi_{m\ell}(y - y_o)^m (z - y_o)^\ell.$$

The assumptions of the theorem imply that

$$\tilde{A} \in G\ell(n;b_{g,\alpha,\rho'}),\ \tilde{\varphi}_{oo} \in (b_{g,\alpha,\rho'})^n,\ \tilde{\psi} \in (\hat{V}^{(2)}_{g,\alpha,\rho'}(y_o))^n,$$

where $\rho' = \rho + \frac{1}{g}(1 - k + \frac{1}{p})$. (cf. formula (16.4) and the corollary to lemma 16.13). Further, we define

$$B = [D_2\varphi(y_o,y_o)],$$

and

$$\Delta_{\pm} = [s^{1-k}]B\ \delta_{\pm} + \tilde{A}.$$

Without loss of generality we may assume that $y_o = 0$. Let $\alpha' \geq \alpha$ and let $U_{\alpha'}(\delta)$ denote the set of all $\eta \in (b_{g,\alpha',\rho'})^n$ such that $\|\eta\|_{g,\alpha',\rho'} \leq \delta$. In view of proposition 16.15 and lemma 16.25 we can define a mapping T on $U_{\alpha'}(\delta)$ by

$$T\eta = -\Delta_{\pm}^{-1}\{\tilde{\varphi}_{oo} + \tilde{\psi}(\eta, \tau_{\pm}\eta)\}, \tag{16.34}$$

provided α' is sufficiently large and $\delta \leq \delta_1$. Applying lemma 16.26 (with $y_1 = \eta$, $y_2 = 0$) we find

$$\|T\eta\|_{g,\alpha',\rho'} \leq \|\Delta_{\pm}^{-1}\|_{g,\alpha',\rho'}\{\|\tilde{\varphi}_{oo}\|_{g,\alpha',\rho'} + D\ \|\eta\|^2_{g,\alpha',\rho'}\} \tag{16.35}$$

for all $\eta \in U_{\alpha'}(\delta)$, provided $\delta \leq \delta_2$. Since $\tilde{A}(0)$ is invertible we have

$$\Delta_{\pm}^{-1} = \{I + \tilde{A}(0)^{-1}(\Delta_{\pm} - \tilde{A}(0))\}^{-1}\tilde{A}(0)^{-1}.$$

Noting that

$$\Delta_{\pm} - \tilde{A}(0) = [s^{1-k}]B\ \delta_{\pm} + \tilde{A} - \tilde{A}(0)$$

and using (16.6) and (16.7) one easily verifies that

$$\lim_{\alpha' \to \infty} \|\Delta_{\pm} - \widetilde{A}(0)\|_{g,\alpha',\rho'} = 0$$

Hence there exist positive constants d and α_o such that

$$\|\Delta_{\pm}^{-1}\|_{g,\alpha',\rho'} \leq d \tag{16.36}$$

for all $\alpha' \geq \alpha_o$. Now, let δ be chosen in such a way that

$$\delta D d < 1.$$

Assumption (ii) of the theorem implies that $\widetilde{\varphi}_{oo} = 0$. Thus, by taking α' sufficiently large, we can achieve that

$$\|\widetilde{\varphi}_{oo}\|_{g,\alpha',\rho'} < (d^{-1} - D\delta)\,\delta. \tag{16.37}$$

Combining (16.35),(16.36) and (16.37) we conclude that

$$\|T\eta\|_{g,\alpha',\rho'} \leq \delta$$

for all $\eta \in U_{\alpha'}(\delta)$, provided α' is sufficiently large.

Moreover, with the aid of (16.34),(16.36) and lemma 16.26 we find that T is a contraction on $U_{\alpha'}(\delta)$. Consequently, there is a unique element η of $U_{\alpha'}(\delta)$ with the property that

$$\eta = -\Delta_{\pm}^{-1}\{\widetilde{\varphi}_{oo} + \widetilde{\psi}(\eta,\tau_{\pm}\eta)\},$$

and thus

$$[s^{k-1}]\Delta_{\pm}\,\eta = -\{[\varphi_{oo}] + [\psi](\eta,\tau_{\pm}\eta)\}.$$

Let y be the element of $(\mathbb{C}[\![s^{-1/p}]\!])^n$ determined by η. Obviously, y is a solution of (16.33) and $y \in (\mathbb{C}[\![s^{-1/p}]\!]_{pg})^n$. □

Finally, we consider the case that $k = 0$.

<u>THEOREM</u> 16.38. *Let* $g \in [0,1]$, $y_o \in \mathbb{C}^n$ *and let* $\varphi \in (\hat{V}_g^{(2)}(y_o))^n$. *Set*

$$D_1\varphi(y_o,y_o) + D_2\varphi(y_o,y_o) = A, \qquad D_2\varphi(y_o,y_o) = B,$$

and assume that

(i) $B \in G\ell(n;\mathbb{C}[\![s^{-1/p}]\!])$,

(ii) $sA \mp hB \in G\ell(n;\mathbb{C}[\![s^{-1/p}]\!])$ *for all* $h \in \frac{1}{p}\mathbb{N}'$,

(iii) $\varphi_{m\ell} \in s^{-1-1/p}(\mathbb{C}[\![s^{-1/p}]\!])^n$ *for all* m,ℓ *such that* $|m+\ell| \neq 1$.

Then equation (16.33) *possesses a unique solution* $y \in (\mathbb{C}[\![s^{-1/p}]\!]_{pg})^n$, *and* $y(\infty) = y_o$.

PROOF: Without loss of generality we may assume that $B = I$ (this can always be achieved by multiplying both sides of (16.33) by $(D_2\varphi(y_o,y_o))^{-1}$). Suppose that $g > 0$. We define a difference operator $\Delta_\pm$ as follows

$$\Delta_\pm = [s]\delta_\pm + \tilde{A},$$

where $\tilde{A} = [sA]$. For suitable values of α and ρ this mapping is a bijection from $(b_{g,\alpha,\rho})^n$ onto $(b_{g,\alpha,\rho+1})^n$. The proof now proceeds analogously to that of theorem 16.32. The only modification that is not quite self-evident occurs in the derivation of (16.36). We may write

$$\Delta_\pm^{-1} = \{I + L^{-1}(\Delta_\pm - L)\}^{-1} L^{-1},$$

where L is the linear mapping defined by

$$La(h) = (\tilde{A}(0) \mp hI)\,a(h), \qquad h \in \frac{1}{p}\mathbb{N}'.$$

In view of the second assumption of theorem 16.38, L is a bijection from $(b_{g,\alpha,\rho})^n$ onto $(b_{g,\alpha,\rho+1})^n$. Obviously, the norm of L^{-1} remains bounded when α tends to infinity. With the aid of formulas (16.4), (16.19) and (16.20) one easily verifies that

$$\lim_{\alpha\to\infty} \sup_{\substack{y \in (b_{g,\alpha,\rho})^n \\ \|y\|_{g,\alpha,\rho} \le 1}} \|(\Delta_\pm - L)y\|_{g,\alpha,\rho+1} = 0.$$

Hence there exist positive constants d and α_o, such that

$$\|\Delta_\pm^{-1} y\|_{g,\alpha',\rho} \le d\, \|y\|_{g,\alpha',\rho+1}$$

for all $\alpha' \ge \alpha_o$ and all $y \in (b_{g,\alpha',\rho})^n$. □

§17. Block-diagonalization and - triangularization theorems.

Let S be a sector (in some cases we shall consider the set $\mathcal{S}$ mentioned in theorem 15.16 instead), let $g \in [0,1]$ and $A \in M_g(S)$. Then A admits an asymptotic expansion of the form

$$\hat{A}(s) = \sum_{h=h_o}^{\infty} A_h \, s^{-h/p}, \qquad h_o \in \mathbb{Z}.$$

Suppose there is an integer $r \in \{0,..,p\}$ such that, for all $h < h_o + r$,

$$A_h = \lambda_h I, \qquad \lambda_h \in \mathbb{C}, \; \lambda_{h_o} \neq 0 \text{ if } r > 0,$$

whereas

$$A_{h_o+r} = \mathrm{diag}\{A_{11}^r, A_{22}^r\},$$

where A_{11}^r and A_{22}^r are square matrices (respectively of order n_1 and n_2) possessing no common eigenvalues. In the case that $r = p$ we shall assume that no two eigenvalues $\tilde{\lambda}_1$ of A_{11}^p and $\tilde{\lambda}_2$ of A_{22}^p differ by a multiple of $\frac{1}{p}\lambda_{h_o}$.

Let $\Delta_\pm$ be the (right or left) difference operator defined by

$$\Delta_\pm y(s) = y(s \pm 1) - A(s)\, y(s).$$

We seek a matrix function $F \in \mathrm{End}(n; A_g(S))$ such that the transformation

$$y(s) = F(s) z(s)$$

changes $\Delta_\pm$ into a difference operator $\Delta_\pm^F$ of the form

$$\Delta_\pm^F z(s) = z(s \pm 1) - B(s)\, z(s),$$

where B has the following properties:

$$B(s) = \mathrm{diag}\{B_{11}(s), B_{22}(s)\}$$

and

$$B_h = A_h \qquad \text{for all } h \in \{h_o,..,h_o + r\}.$$

F must satisfy the equation

$$A(s)F(s) = F(s \pm 1)B(s). \tag{17.1}$$

We put

$$A(s) = \begin{pmatrix} A_{11}(s) & A_{12}(s) \\ A_{21}(s) & A_{22}(s) \end{pmatrix}$$

and take F to be of the form

$$F(s) = \begin{pmatrix} I_{n_1} & F_{12}(s) \\ F_{21}(s) & I_{n_2} \end{pmatrix}. \tag{17.2}$$

Equating corresponding blocks on both sides of (17.1) and eliminating B_{11} and B_{22} we obtain

$$F_{ij}(s \pm 1)\{A_{jj}(s) + A_{ji}(s)F_{ij}(s)\} = A_{ij}(s) + A_{ii}(s)F_{ij}(s), \quad (i,j) = (1,2) \text{ or } (2,1).$$

Let the function Φ_{ij} be defined by

$$\Phi_{ij}(s,y,z) = A_{ij}(s) + A_{ii}(s)y - zA_{jj}(s) - zA_{ji}(s)y, \tag{17.3}$$

where $y,z \in \mathrm{Hom}(\mathbb{C}^{n_j},\mathbb{C}^{n_i})$, $(i,j) = (1,2)$ or $(2,1)$. We are looking for solutions of the equations

$$\Phi_{ij}(s,y(s),y(s \pm 1)) = 0, \qquad (i,j) = (1,2) \text{ or } (2,1). \tag{17.4}$$

In view of the isomorphism existing between the linear spaces $\mathrm{Hom}(\mathbb{C}^{n_j},\mathbb{C}^{n_i})$ and $\mathbb{C}^{n_1 n_2}$, we may identify $n_i \times n_j$-matrices with $n_1 n_2$-dimensional vectors. Doing this we find that

$$\Phi_{ij} \in (V_g^{(2)}(S,0))^{n_1 n_2} \qquad (i,j) = (1,2) \text{ or } (2,1).$$

It may happen that we are able to solve the equation (17.4) for one pair $(i,j) \in \{(1,2),(2,1)\}$, but not for the other. In that case we can put the matrix A into a block triangular form. If F_{12} is a solution of (17.4) with $i = 1$, $j = 2$, we get

$$\begin{pmatrix} I_{n_1} & F_{12}(s\pm 1) \\ 0 & I_{n_2} \end{pmatrix}^{-1} A(s) \begin{pmatrix} I_{n_1} & F_{12}(s) \\ 0 & I_{n_2} \end{pmatrix} =$$

$$= \begin{pmatrix} A_{11}(s)-F_{12}(s\pm 1)A_{21}(s) & 0 \\ A_{21}(s) & A_{21}(s)F_{12}(s)+A_{22}(s) \end{pmatrix}. \tag{17.5}$$

If F_{21} is a solution of (17.4) with $i=2$, $j=1$, we have

$$\begin{pmatrix} I_{n_1} & 0 \\ F_{21}(s\pm 1) & I_{n_2} \end{pmatrix}^{-1} A(s) \begin{pmatrix} I_{n_1} & 0 \\ F_{21}(s) & I_{n_2} \end{pmatrix} =$$

$$= \begin{pmatrix} A_{11}(s)+A_{12}(s)F_{21}(s) & A_{12}(s) \\ 0 & A_{22}(s)-F_{21}(s\pm 1)A_{12}(s) \end{pmatrix}. \tag{17.6}$$

LEMMA 17.7. *Let* $(i,j) = (1,2)$ *or* $(2,1)$ *and let* Φ_{ij} *be defined by* (17.3). *In addition to the assumptions made on* p. 108 *suppose that*

$$g \leq 1-\frac{r}{p} \quad \text{if} \quad r<p.$$

Then equation (17.4) *possesses a unique formal solution* $Y_{ij} \in s^{-1/p}\,\mathrm{Hom}((\mathbb{C}[\![s^{-1/p}]\!]_{pg})^{n_j}, (\mathbb{C}[\![s^{-1/p}]\!]_{pg})^{n_i})$.

PROOF: We may assume that $h_o = 0$ (If not, we multiply both sides of (17.4) by the factor $s^{h_o/p}$). Considered as a vector function, $\hat{\Phi}_{ij}$ clearly belongs to $(\hat{V}_g^{(2)}(0))^{n_1 n_2}$. Let $k = 1-\frac{r}{p}$. According to the assumptions made on p.108, $A_{ij}(s) = O(s^{k-1-1/p})$ as $|s| \to \infty$ in S. Hence it follows that the second condition of theorem 16.32, or, if $k=0$, the third condition of theorem 16.38, is fulfilled. Further, note that $D_2\hat{\Phi}_{ij}(s,0,0)$ is the linear mapping

$$x \longrightarrow \hat{A}_{ii}x, \qquad x \in \mathrm{Hom}(\mathbb{C}^{n_j}, \mathbb{C}^{n_i}),$$

and $D_3\hat{\Phi}_{ij}(s,0,0)$ is the linear mapping

$$x \longrightarrow -x\hat{A}_{jj}, \qquad x \in \mathrm{Hom}(\mathbb{C}^{n_j}, \mathbb{C}^{n_i}).$$

Let us put

$$D_2\hat{\Phi}_{ij}(s,0,0) + D_3\hat{\Phi}_{ij}(s,0,0) = \sum_{h\in\frac{1}{p}\mathbb{N}'} A^{ij}(h)\, s^{-h},$$

where the $A^{ij}(h)$ are $n_1n_2 \times n_1n_2$ matrices. Obviously, $A^{ij}(h) = 0$ for all $h < 1-k$. All eigenvalues of $A^{ij}(1-k)$ can be obtained as the difference of an eigenvalue of A^r_{ii} and one of A^r_{jj} and are therefore unequal to 0, and, in the case that $k = 0$, to any multiple of $\frac{1}{p}\lambda_o$. Furthermore, we note that $D_2\hat{\Phi}_{ij}(\infty,0,0) = \lambda_o I$ if $r > 0$ and thus in particular if $k = 0$. Apparently, all conditions of theorem 16.32 (if $k > 0$) or 16.38 (if $k = 0$) are satisfied. Consequently, equation (17.4) has a unique formal solution $Y_{ij} \in (\mathbb{C}[\![s^{-1/p}]\!]_{pg})^{n_1n_2}$ with the property that $Y_{ij}(\infty) = 0$. Hence the conclusion of the lemma follows immediately. □

From now on we shall use the following notation. We put

$$1 - \frac{r}{p} = k.$$

Let $i \in \{1,2\}$. By

$$\lambda_{i,h}\ ,\ h = 1,\ldots,\ell_i \quad (\ell_i \in \mathbb{N})$$

we shall denote:

a) the eigenvalues of A^r_{ii} if $r \neq 0$,

b) the non-zero eigenvalues of A^o_{ii} if $r = 0$.

Finally, let $(i,j) = (1,2)$ or $(2,1)$. $\tilde{A}_{ij}$ will denote the $n_1n_2 \times n_1n_2$ matrix function defined by

$$\tilde{A}_{ij}(s) = -D_3\hat{\Phi}_{ij}(s,Y_{ij}(s),Y_{ij}(s\pm1))^{-1} D_2\hat{\Phi}_{ij}(s,Y_{ij}(s),Y_{ij}(s\pm1)), \tag{17.8}$$

where Y_{ij} is a formal solution of (17.4).

<u>LEMMA</u> 17.9. *Let* $(i,j) = (1,2)$ or $(2,1)$. *In addition to the assumptions made on* p.108 *suppose that* $k = 0$ *or* $k \geq g$. *Then* $\tilde{A}_{ij} \in G\ell(n_1n_2;\mathbb{C}[\![s^{-1/p}]\!]_{pg}[s^{1/p}])$. *Moreover, if* A_{h_o} *is invertible, we have that* $\tilde{A}_{ij} \in G\ell(n_1n_2;\mathbb{C}[\![s^{-1/p}]\!]_{pg})$. *If* $k = 1$ *and* A^o_{jj} *is invertible, then* $\tilde{A}_{ij} \in \mathrm{End}(n_1n_2;\mathbb{C}[\![s^{-1/p}]\!])$, *and if* A^o_{ii} *is invertible, then* $\tilde{A}^{-1}_{ij} \in \mathrm{End}(n_1n_2;\mathbb{C}[\![s^{-1/p}]\!])$. *Furthermore, if* $k < 1$, *or if* $k = 1$ *and neither* A^o_{11} *nor* A^o_{22} *is nilpotent, the following two statements hold*

(i) $k(\widetilde{A}_{ij}) = \{k\}$,

(ii) $\mu \in \mu_k^{\pm}(\widetilde{A}_{ij})$ *if and only if there exist integers* $h_1 \in \{1,..,\ell_1\}$ *and* $h_2 \in \{1,..,\ell_2\}$ *such that*

$$\mu = \pm \frac{1}{k} \lambda h_o^{-1} (\lambda_{i,h_i} - \lambda_{j,h_j}) \qquad \text{if } k < 1, \tag{17.10}$$

or

$$\mu = \pm \log(\lambda_{i,h_i} \lambda_{j,h_j}^{-1}) \qquad \text{if } k = 1. \tag{17.11}$$

In the latter relation all determinations of the logarithm are allowed.

<u>PROOF:</u> $D_3\hat{\Phi}_{ij}(s,Y_{ij}(s),Y_{ij}(s\pm 1))$ is the linear mapping

$$x \longrightarrow -x\,(\hat{A}_{jj}(s) + \hat{A}_{ji}(s)Y_{ij}(s)), \qquad x \in \mathrm{Hom}(\mathbb{C}^{nj},\mathbb{C}^{ni}),$$

while $D_2\hat{\Phi}_{ij}(s,\ Y_{ij}(s),Y_{ij}(s\pm 1))$ is the mapping

$$x \longrightarrow (\hat{A}_{ii}(s) - Y_{ij}(s\pm 1)\hat{A}_{ji}(s))\,x\,, \qquad x \in \mathrm{Hom}(\mathbb{C}^{nj},\mathbb{C}^{ni}).$$

Note that

$$\hat{A}_{jj}(s) + \hat{A}_{ji}(s)Y_{ij}(s) = \hat{B}_{jj}(s),$$

where B_{jj} has been defined on p.108. Let us put

$$\hat{A}_{ii}(s) - Y_{ij}(s\pm 1)\hat{A}_{ji}(s) = \widetilde{B}_{ii}(s).$$

By assumption, $A \in G\ell(n;\mathbb{C}[\![s^{-1/p}]\!]_{pg}[s^{1/p}])$. Further, we know that $Y_{ij} \in s^{-1/p}\,\mathrm{Hom}((\mathbb{C}[\![s^{-1/p}]\!]_{pg})^{nj},(\mathbb{C}[\![s^{-1/p}]\!]_{pg})^{ni})$. With the aid of formulas (17.5) and (17.6) we conclude that $\hat{B}_{jj} \in G\ell(n_j;\mathbb{C}[\![s^{-1/p}]\!]_{pg}[s^{1/p}])$ and $\widetilde{B}_{ii} \in G\ell(n_i;\mathbb{C}[\![s^{-1/p}]\!]_{pg}[s^{1/p}])$. Hence the first assertion of the lemma follows.

We may write $\hat{B}_{jj}$ in the form

$$\hat{B}_{jj}(s) = s^{-h_o/p}\{\sum_{\ell=o}^{r-1} \lambda_{h_o+\ell}\, s^{-\ell/p}\, I_{n_j} + s^{-r/p} A_{jj}^{r} + s^{-(r+1)/p}\, C(s)\} \qquad \text{if } r > 0,$$

or

$$\hat{B}_{jj}(s) = s^{-h_o/p}\{A_{jj}^{o} + s^{-1/p}\, C(s)\} \qquad \text{if } r = 0,$$

where $C \in \mathrm{End}(n_j;\mathbb{C}[\![s^{-1/p}]\!]_{pg})$. It can be shown that a transformation of $\hat{B}_{jj}$ into a canonical form does not affect the eigenvalues of the matrix A^r_{jj} (cf. [39]). Similar considerations hold with respect to the matrix $\tilde{B}_{ii}$. Let $\tilde{B}^c_{ii}$ and B^c_{jj} be right or left canonical forms of $\tilde{B}_{ii}$ and B_{jj}, respectively. One easily verifies that the matrix function corresponding to the linear mapping

$$x \longrightarrow B^c_{ii}\, x\, (B^c_{jj})^{-1}, \qquad x \in \mathrm{Hom}(\mathbb{C}^{n_j},\mathbb{C}^{n_i}),$$

is a right or left canonical form of $\tilde{A}_{ij}$. The remaining statements of the lemma now follow easily. □

Application of theorems 15.1 and 15.16 (in the generalized form discussed at the end of section 15) to equation (17.4) yields the following two block diagonalization theorems.

THEOREM 17.12. *Let* $h_o \in \mathbb{Z}$, $r \in \{0,..,p\}$, $g \in [0,1]$ *if* $r = p$, $g \in [0,1-\frac{r}{p}]$ *if* $r < p$. *Let*

$$A \in s^{-h_o/p}\, G\ell(n;A_g(S_o)),$$

where $S_o = S(\alpha_1^o,\alpha_2^o)$ *and* $\pm S_o$ *is a right sector. In addition to the assumptions made on* p.108 *assume that either* $k(\alpha_2^o - \alpha_1^o) \leq \pi$, *or else that*

(i) if $k(\alpha_2^o - \alpha_1^o) > \pi$, then

$$(\alpha_1^o,\alpha_2^o - \tfrac{\pi}{k}) \cap \Sigma_k^{\pm}(A_{ij}) = \phi \quad \text{for } (i,j) = (1,2) \text{ and for } (i,j) = (2,1),$$

(ii) if $k = 1$ and $\alpha_2^o - \alpha_1^o < \pi$, then $[\alpha_2^o - \pi,\alpha_1^o]$ contains at most one element of $\Sigma^{\pm}_{1,h}(\tilde{A}_{ij})$ for all $h \in \{1,..,m\}$, where m is the number of blocks in the canonical form of $\tilde{A}_{ij}$.

Here $k = 1-\frac{r}{p}$, $\tilde{A}_{ij}$ *is defined by* (17.8). *Then there exists a matrix function* $F \in G\ell(n;A_g(S_o))$ *of the form* (17.2) *such that the transformed matrix*

$$F(s\pm1)^{-1}\, A(s)F(s)$$

is block diagonal in the same partition as A_{h_o+r}. *Moreover, the matrix function with these properties is unique if* $k = 0$ *or* $k(\alpha_2^o - \alpha_1^o) > \pi$, *or, in the case that* $0 < k(\alpha_2^o - \alpha_1^o) \leq \pi$ *if the additional condition*

$$[\alpha_2^o - \tfrac{\pi}{k},\alpha_1^o] \cap \Sigma_k^{\pm}(\tilde{A}_{ij}) = \phi \quad \textit{for } (i,j) = (1,2) \textit{ and for } (i,j) = (2,1)$$

is satisfied.

REMARK. The k-singular directions of $\tilde{A}_{ij}$ can be calculated from the eigenvalues of A^r_{ii} and A^r_{jj} with the aid of lemma 17.9.

THEOREM 17.13. *Let* $S_o = S(\alpha^o_1, \alpha^o_2)$, *with* $\alpha^o_1 = \mp\frac{\pi}{2}$ *and* $\alpha^o_2 = \alpha^o_1 + \pi$. *Let* $A \in M_g(\mathbf{S})$, *where* $g \in [0,1)$ *and* $\mathbf{S}$ *is a set of* S_o*-proper regions as defined in* §12.1. *Let* $\hat{A}(s) = \sum_{h=h_o}^{\infty} A_h s^{-h/p}$ *and assume that*

$$A_{h_o} = \mathrm{diag}\{A^o_{11}, A^o_{22}\},$$

where A^o_{11} *is a nilpotent* $n_1 \times n_1$ *matrix,* A^o_{22} *is an invertible* $n_2 \times n_2$ *matrix. Then there exists a unique matrix function* $F \in G\ell(n;A_g(\mathbf{S}))$ *of the form* (17.2) *, such that the transformed matrix*

$$F(s\pm1)^{-1}\, A(s)\, F(s)$$

is block diagonal in the same partition as A_{h_o}.

As we pointed out before, we may put the matrix function A into a block triangular form by solving equation (17.4) for one pair (i,j) and performing the transformation (17.5) or (17.6)(according to whether (i,j) = (1,2) or (2,1)). With the aid of theorems 15.1 , 15.12 and 15.15 and lemmas 17.7 and 17.9 we obtain the following results.

THEOREM 17.14 *Let* $r \in \{0,..,p\}, g \in [0,1]$ *if* $k = 0$, $g \in [0,k]$ *if* $k > 0$. $(k = 1 - \frac{r}{p})$. *Let* $A \in M_g(S_o)$, *where* $S_o = S(\alpha^o_1, \alpha^o_2) = \pm\, S(-\pi,\pi)$.

Let (i,j) = (1,2) *or* (2,1). *We make the following assumptions in addition to those mentioned on* p.108,

(i) *if* $\frac{1}{2} < k < 1$: *corresponding to each pair* $(h_1,h_2) \in \{1,..,\ell_1\} \times \{1,..,\ell_2\}$ *there exists a real number* $\alpha \in [\alpha^o_2 - \frac{2\pi}{k}, \alpha^o_1]$ *such that*

$$\pm\mathrm{Re}\{\lambda^{-1}_{h_o}(\lambda_{i,h_i} - \lambda_{j,h_j})e^{ik\alpha}\} \leq 0. \tag{17.15}$$

(ii) *if* $k = 1$: A^o_{ii} *is invertible and, furthermore,*

$$|\lambda_{i,h_i}\, \lambda^{-1}_{j,h_j}| \geq 1 \tag{17.16}$$

for all $h_1 \in \{1,..,\ell_1\}$ *and all* $h_2 \in \{1,..,\ell_2\}$.
Then there exists a matrix function $F \in G\ell(n;A_g(S_o))$ *of the form* $\begin{pmatrix} I_{n_1} & F_{12} \\ 0 & I_{n_2} \end{pmatrix}$ *if* $(i,j) = (1,2)$, *or* $\begin{pmatrix} I_{n_1} & 0 \\ F_{21} & I_{n_2} \end{pmatrix}$ *if* $(i,j) = (2,1)$, *such that the transformed matrix*

$$F(s\pm 1)^{-1}A(s)F(s)$$

is block triangular in the same partition as A_{h_o+r} . *Moreover, if* $k=0$ *or* $k>\frac{1}{2}$ *the matrix function with these properties is unique.*

PROOF: We shall consider the case that S_o is a right sector. First, suppose that $k=1$. If A^o_{ii} is invertible, then $\widetilde{A}^{-1}_{ij} \in \mathrm{End}(n_i \times n_j; \mathbb{C}[[s^{-1/p}]]_{pg})$, where $\widetilde{A}_{ij}$ is the matrix function defined in (17.8). Furthermore, (17.16) implies that $\arg\{\log(\lambda_{i,h_i}\lambda^{-1}_{j,h_j})\} \in [-\frac{\pi}{2},\frac{\pi}{2}] \bmod 2\pi$. Hence we deduce, using (17.11), that $\Sigma^+_1(\widetilde{A}_{ij}) \subset [0,\pi] \bmod 2\pi$, which shows that $(-\pi,0) \cap \Sigma^+_1(\widetilde{A}_{ij}) = \emptyset$. Thus, conditions (i) and (ii) of theorem 15.1 are satisfied. If $k<1$, then $A_{h_o} = I$ and, consequently, $\widetilde{A}_{ij} \in G\ell(n_i \times n_j; \mathbb{C}[[s^{-1/p}]]_{pg})$. Suppose that $k>\frac{1}{2}$. (17.15) is seen to imply that $\frac{1}{k}[\frac{\pi}{2} - \arg\{\lambda^{-1}_{h_o}(\lambda_{i,h_i} - \lambda_{j,h_j})\}] \in [\alpha - \frac{\pi}{k}, \alpha] \bmod \frac{2\pi}{k}$. With (17.10) it follows that $\Sigma^+_k(\widetilde{A}_{ij}) \subset [\alpha^o_2 - \frac{3\pi}{k}, \alpha^o_1] \bmod \frac{2\pi}{k}$. If, on the other hand, $k \leq \frac{1}{2}$, then we have that $k(\alpha^o_2 - \alpha^o_1) \leq \pi$. Obviously, in both cases conditions (i), (ii) and (iii) of theorem 15.1 are fulfilled. In view of lemma 17.7 we now conclude that all conditions of theorem 15.1 are satisfied. Application of that theorem yields the desired result. □

THEOREM 17.17. *Let* $r \in \{0,..,p-1\}$, $0 \leq g \leq 1 - \frac{r}{p}$ $(=k)$. *Let* $S_o = S(\alpha^o_1,\alpha^o_2) = \pm S(-\pi,\pi)$ *and* $A \in M_g(S_t)$, *where* $t \in \{1,2\}$ *and* S_t *is defined by* (15.11). *Let* $(i,j) = (1,2)$ *or* $(2,1)$. *We make the following assumptions in addition to those on* p. 108 ,

(i) *if* $k=1$, *then* A^o_{ii} *is invertible.*

(ii) *for all* $h_1 \in \{1,..,\ell_1\}$ *and all* $h_2 \in \{1,..,\ell_2\}$ *the following condition is satisfied*

$$\pm\mathrm{Re}\{\lambda^{-1}_{h_o}(\lambda_{i,h_i} - \lambda_{j,h_j})e^{ik\alpha^o_t}\} < 0 \qquad \text{if } k<1$$

and

$$|\lambda_{i,h_i}\lambda^{-1}_{j,h_j}| > 1 \qquad\qquad \text{if } k=1.$$

Then there exists a matrix function $F \in G\ell(n;A_g(S_t))$, *of the form* $\begin{pmatrix} I_{n_1} & F_{12} \\ 0 & I_{n_2} \end{pmatrix}$ *if* $(i,j) = (1,2)$, *or* $\begin{pmatrix} I_{n_1} & 0 \\ F_{21} & I_{n_2} \end{pmatrix}$ *if* $(i,j) = (2,1)$, *such that the transformed matrix*

$$F(s\pm1)^{-1}A(s)F(s)$$

is block triangular in the same partition as A_{h_o+r} *Moreover, if* $k > \frac{1}{2}$, *then the matrix function with these properties is unique.*

The above result can be easily proved by appealing to theorem 15.12. Application of theorem 15.15 yields the following result for upper or lower half planes.

<u>THEOREM</u> 17.18 *Let* $r \in \{0,..,p\}, g \in [0,1]$ *if* $k=0, g \in [0,k]$ *if* $k>0$ $(k = 1-\frac{r}{p})$. *Let* $t \in \{1,2\}$ *and let* α_1^o *and* α_2^o *be real numbers such that* $\alpha_t^o = \ell\pi, \ell \in \mathbb{Z}$ *and* $\alpha_2^o - \alpha_1^o = \pi$. *Suppose that* $A \in M_g(S_t)$, *where* S_t *is defined by* (15.11). *Let* $(i,j) = (1,2)$ *or* $(2,1)$. *We make the following assumptions in addition to those mentioned on* p. 108 ,

(i) *if* $k=1$, *then* A^o_{ii} *is invertible.*

(ii) *for all* $h_1 \in \{1,..,\ell_1\}$ *and all* $h_2 \in \{1,..,\ell_2\}$ *the following condition is satisfied*

$$\pm\mathrm{Re}\{\lambda^{-1}_{h_o}(\lambda_{i,h_i} - \lambda_{j,h_j})e^{ik\alpha_t^o}\} > 0 \qquad\qquad \text{if } 0<k<1,$$

and

$$|\lambda_{i,h_i}\lambda^{-1}_{j,h_j}| > 1 \qquad\qquad \text{if } k=1.$$

Then there exists a unique matrix function $F \in G\ell(n;A_g(S_t))$, *of the form* $\begin{pmatrix} I_{n_1} & F_{12} \\ 0 & I_{n_2} \end{pmatrix}$ *if* $(i,j) = (1,2)$, *or* $\begin{pmatrix} I_{n_1} & 0 \\ F_{21} & I_{n_2} \end{pmatrix}$ *if* $(i,j) = (2,1)$, *such that the transformed matrix*

$$F(s\pm 1)^{-1}A(s)F(s)$$

is block triangular in the same partition as A_{h_0+r}. *(We use the upper sign if* ℓ *is even, the lower sign if* ℓ *is odd).*

REMARK The theorems 17.14, 17.17 and 17.18 can be generalized in an obvious manner to sectors of the type considered in theorems 15.1, 15.12 and 15.15, respectively.

§18. Application to canonical forms.

We shall now apply the foregoing results to the study of the homogeneous linear difference equation

$$y(s+1) = A(s)y(s), \tag{18.1}$$

where A is an n by n matrix function, holomorphic in a set S of the form $S = \{s \in \mathbb{C}: \alpha_1 < \arg s < \alpha_2, |s| > R\}$, and represented asymptotically by a power series in $s^{-1/p}$ $(p \in \mathbb{N})$ as $|s| \to \infty$ in S. In particular, one may think of A as being meromorphic at infinity, which is the most interesting case. If, for all $s \in S$, $A(s-1)^{-1}$ exists, then (18.1) is equivalent to the equation

$$y(s-1) = A(s-1)^{-1}y(s). \tag{18.2}$$

All results concerning the first class of equations can be easily translated into results for the second class and vice versa. One readily verifies that the matrix functions $A(s)^{-1}$ and $A(s-1)^{-1}$ have identical (left and right) canonical forms. From §2 we recall the relations

$$d^-(A^{-1}) = d^+(A), k(A^{-1}) = k(A), \mu_k^-(A^{-1}) = \mu_k^+(A) \quad \text{for all } k \in k(A), \tag{18.3}$$

and

$$\gamma^-(A^{-1}) = \gamma^+(A).$$

Let $A^c = A_+^c$ be a right canonical form of A as defined in (2.3) - (2.5). We shall prove the existence, under quite general conditions, of a fundamental matrix Y of (18.1) of the form

$$Y(s) = F(s)\, s^{Ds}\, e^{Q(s)}\, s^{G},$$

where F is a holomorphic matrix function in S, admitting an asymptotic expansion $\hat{F} \in G\ell(n;\mathbb{C}[\![s^{-1/p}]\!][s^{1/p}])$. D,Q and G have been defined in (2.3) - (2.5). Alternatively, the problem considered here may be formulated in the following manner: If A^c is a (right) canonical form of A, find a matrix function $F \in G\ell(n;A_o(S(\alpha_1,\alpha_2))[s^{1/p}])$ such that

$$A^c(s) = F(s+1)^{-1}A(s)F(s).$$

Obviously, F must satisfy the equation

$$F(s+1) = A(s)F(s)A^c(s)^{-1}. \tag{18.4}$$

<u>DEFINITION</u> *Let* A *be an invertible square matrix. By* $\sigma(A)$ *we shall denote a matrix representing the linear mapping*

$$x \to A\,x\,A^{-1}$$

If A *is a matrix function,* $\sigma(A)$ *will denote the function defined by*

$$\sigma(A)(s) = \sigma(A(s)).$$

One easily verifies that $\sigma(A) = A \otimes (A^{-1})^T$ ($(A^{-1})^T$ is the transpose of A^{-1}). If A^c is a canonical form of A, then $\sigma(A^c)$ is a canonical form of $\sigma(A)$. Hence we deduce the following statements.

<u>LEMMA</u> 18.5. *Let* $A \in \hat{M}_o$ *and let the matrix function* $A^c = A^c_+$ *defined in* (2.3) - (2.5) *be a right canonical form of* A. *Then the following relations hold.*

(i) $d^+(\sigma(A)) = \{d_i - d_j;\ i,j \in \{1,..,m\}\}$.

(ii) $\gamma \in \gamma^+(\sigma(A))$ *iff there are integers* $i,j \in \{1,..,m\}$ *such that* $\gamma - (\gamma_i - \gamma_j) \in \frac{1}{p}\mathbb{Z}$ *and* $\operatorname{Re}\gamma \in [0,\frac{1}{p})$.

(iii) $k \in k(\sigma(A))$ *iff there are integers* $i,j \in \{1,..,m\}$ *such that* $d_i = d_j$ *and* $\operatorname{degr}(\underline{q_i - q_j}) = k$.

(iv) *Let* $k \in k(\sigma(A))$. $\mu \in \mu_k^+(\sigma(A))$ *iff there are integers* $i,j \in \{1,..,m\}$ *such that* $d_i = d_j$, $\operatorname{degr}(\underline{q_i - q_j}) = k$ *and* $\mu_{i,k} - \mu_{j,k} = \mu$.

The most satisfactory results are obtained in the case that $\hat{A} \in G\ell(n;\mathbb{C}[\![s^{-1/p}]\!])$, or, more generally, $\hat{A} \in s^d G\ell(n;\mathbb{C}[\![s^{-1/p}]\!])$, $d \in \frac{1}{p}\mathbb{Z}$. In fact, the resemblances to the theory of differential equations are strongest in this case.

<u>THEOREM</u> 18.6. *Let* $A \in s^d G\ell(n;A_g(S_o)) \cap M_g(S_o)$, *where* $S_o = S(\alpha_1^o, \alpha_2^o)$ *is a left or right sector,* $d \in \frac{1}{p}\mathbb{Z}$ *and* $g \in [0,1]$. *Let* A^c *be a right canonical form of* A. *Assume that*

(i) $g \leq \min\{k \in k(\sigma(A)) : k > 0\}$

(ii) *for all* $k \in k(\sigma(A))$ *such that* $k(\alpha_2^o - \alpha_1^o) > \pi$ *the following condition is satisfied*

$$(\alpha_1^o, \alpha_2^o - \frac{\pi}{k}) \cap \Sigma_k^+(\sigma(A)) = \phi$$

(iii) *if* $\alpha_2^o - \alpha_1^o < \pi$ *then* $[\alpha_2^o - \pi, \alpha_1^o]$ *contains at most one element of* $\Sigma_{1,j}^+(\sigma(A))$ *for each* $j \in \{1,..,M\}$, *where* M *is the number of blocks in* $\sigma(A^c)$ (*cf. the definition of* $\Sigma_{1,j}^+(A)$ *on p.8*).
Then there exists a matrix function $F \in G\ell(n;A_g(S_o)[s^{1/p}])$ *such that*

$$F(s+1)^{-1}A(s)F(s) = A^c(s). \tag{18.7}$$

<u>PROOF</u>: First we consider the case that S_o is a right sector. Let r_o be a multiple of $\frac{1}{p}$ such that

$$r_o > 1 - d$$

There exists a matrix function $F_o \in G\ell(n;\mathbb{C}\{s^{-1/p}\}[s^{1/p}])$ which transforms A into

$$F_o(s+1)^{-1}A(s)F_o(s) = A^c(s) + s^{-r_o}\tilde{A}(s), \tag{18.8}$$

where $\tilde{A} \in \mathrm{End}(n;A_g(S_o))$ (cf. §2, in particular (2.7)). Putting

$$A^c(s) + s^{-r_o}\tilde{A}(s) = B(s) \tag{18.9}$$

and

$$F(s) = F_o(s)Z(s)$$

and substituting this into (18.4) we find that Z must satisfy the equation

$$Z(s+1) = B(s)Z(s)A^c(s)^{-1}. \tag{18.10}$$

We shall prove the statement made in theorem 18.6 by means of induction on m, the number of blocks in A^c (cf.(2.3)). If $m = 1$, then (18.10) can be rewritten in the form

$$Z(s+1) - \{(1+s^{-1})^N + s^{-1-1/p}C(s)\}Z(s)(1+s^{-1})^{-N} = 0 \,, \tag{18.11}$$

where N is a nilpotent matrix and $C \in \mathrm{End}(n;A_g(S_o))$. Note that the linear mapping

$$Z \longrightarrow ZN - NZ$$

is nilpotent. If $g > 0$ it follows from proposition 16.21 and in the case that $g = 0$ it is a well-known fact, that (18.11) possesses a formal solution $\tilde{F}_1 \in G\ell(n;\mathbb{C}[\![s^{-1/p}]\!]_{pg})$ with the property that $\tilde{F}_1(\infty) = I$. According to theorem 15.1 there exists a matrix function $F_1 \in \mathrm{End}(n;A_g(S_o))$ such that F_1 is a solution of (18.11) and $\hat{F}_1 = \tilde{F}_1$. Obviously, $F_1 \in G\ell(n;A_g(S_o))$. Now, let $m > 1$ and suppose the assertion in theorem 18.6 is true for all $m' < m$. The following two situations may arise:

1. Not all polynomials $\underline{q_j}$ are equal. This implies that $\max k(\sigma(A)) > 0$. Let $\max k(\sigma(A)) = k$. By assumption, $k \geq g$. Without loss of generality we may assume that $\mu_{j,k} = \mu_{1,k}$ for $j = 1,..,m_1$, where $m_1 < m$, and $\mu_{j,k} \neq \mu_{1,k}$ for $j = m_1+1,..,m$. (If $k = 1$, $\mu_{j,k}$ should be replaced by $\underline{\mu_{j,1}}$). Let $N_1 = n_1 + .. + n_{m_1}$ and $N_2 = n - N_1$. Partitioning B along the N_1 th row and column, we may write

$$B = \begin{pmatrix} B_{11} & B_{12} \\ B_{21} & B_{22} \end{pmatrix}.$$

With the aid of lemmas 18.5, 17.9 and the relations (2.6) one easily verifies that the conditions of theorem 17.12 are satisfied. Hence there exists a matrix function $F_1 \in G\ell(n;A_g(S_o))$ such that

$$F_1(s+1)^{-1}B(s)F_1(s) = \tilde{B}(s) = \mathrm{diag}\{\tilde{B}_{11}(s), \tilde{B}_{22}(s)\}, \tag{18.12}$$

where $\tilde{B}_{11} \in s^d G\ell(N_1;A_g(S_o))$, $\tilde{B}_{22} \in s^d G\ell(N_2;A_g(S_o))$. Obviously, A^c is a canonical form of $\tilde{B}$. Since both N_1 and N_2 are smaller than n, we may apply theorem 18.6 to $\tilde{B}_{11}$ and $\tilde{B}_{22}$. Thus we find that the assertion of the theorem is also true for $\tilde{B}$ and hence for A.

2. $k(\sigma(A)) = \{0\}$(i.e. $\underline{q_i} = \underline{q_j}$ for all $i,j \in \{1,..,m\}$), but not all numbers γ_j $(j = 1,..,m)$ are equal. In this case too we can transform B into a block diagonal matrix function of the form (18.12) by application of theorem 17.12, this time with $k = 0$. It follows that the conclusion of theorem 18.6 is true for all $m \in \mathbb{N}$.

Now suppose that S_o is a left sector. Using (18.3) and arguing as above, one can prove the existence of a matrix function $F \in G\ell(n;A_g(S_o)[s^{1/p}])$ such that

$$F(s-1)^{-1}A(s-1)^{-1}F(s) = A^c(s-1)^{-1},$$

which is equivalent to (18.7). □

REMARK. If it is known beforehand that (18.4) possesses a formal solution $\tilde{F} \in G\ell(n;\mathbb{C}[\![s^{-1/p}]\!]_{pg}[s^{1/p}])$, then one can prove the existence of a solution $F \in G\ell(n;A_g(S_o)[s^{1/p}])$ with the property that $\hat{F} = \tilde{F}$ in the following manner. First a preliminary transformation of the form (18.8) is performed, where F_o is chosen in such a way that (18.10) possesses a formal solution $\tilde{F}_1 \in G\ell(n;\mathbb{C}[\![s^{-1/p}]\!]_{pg})$. Then theorem 15.1 can be applied directly to the equation (18.10).

THEOREM 18.13. *Let* $A \in M_g(\overline{S}_o)$, *where* $S_o = S(-\frac{\pi}{2}, \frac{\pi}{2})$ *or* $S(\frac{\pi}{2}, \frac{3\pi}{2})$ *and* $g \in [0,1)$. *Let* A^c *be a right canonical form of* A. *If*

$$g \leq \min\{k \in k(\sigma(A)) : k > 0\}$$

then there exists a matrix function $F \in G\ell(n;A_g(S_o)[s^{1/p}])$ *such that*

$$F(s+1)^{-1}A(s)F(s) = A^c(s).$$

PROOF: This theorem can be proved by means of induction on the number of distinct elements of $d^+(A)$. We shall restrict ourselves to a discussion of the case that S_o is a right sector. Further, we assume that $\max d^+(A) = 0$. (If this is not the case, we may multiply A by a suitable power of s since this does not affect the conclusions of the theorem). If $d^+(A) = \{0\}$ the theorem is a special case of theorem 18.6. So suppose that not all d_j $(j = 1,..,m)$ are equal. Without loss of generality we may assume that $d_j < 0$ for $j = 1,..,m_1$, where $m_1 < m$ and $d_j = 0$ for $j = m_1 + 1,..,m$. Let $N_1 = n_1 + .. + n_{m_1}$ and $N_2 = n - N_1$. We begin by performing a preliminary transformation of the form (18.8)[9]. Next, we partition the matrix function B defined in (18.9) along the N_1 th row and column and put

$$B = \begin{pmatrix} B_{11} & B_{12} \\ B_{21} & B_{22} \end{pmatrix}.$$

Then $B_{11}(\infty) = B_{12}(\infty) = B_{21}(\infty) = 0$, whereas $B_{22}(\infty)$ is invertible. Let $\mathcal{S}$ be a set

9) with $r_o > 1 - \min d^+(A)$

of $\overline{S}_o$-proper regions as defined in §12.1. Obviously, $M_g(\mathbf{S})\subset M_g(\overline{S}_o)$. (As a matter of fact, the condition $A\in M_g(\overline{S}_o)$ in theorem 18.13 could be replaced by $A\in M_g(\mathbf{S})$). According to theorem 17.13 there exists a matrix function $F_1\in G\ell(n;A_g(\mathbf{S}))$ such that

$$F_1(s+1)^{-1}B(s)F_1(s)=\mathrm{diag}\{\widetilde{B}_{11}(s),\widetilde{B}_{22}(s)\},$$

where $\widetilde{B}_{11}\in G\ell(N_1;A_g(\mathbf{S})[s^{1/p}])$ and $\widetilde{B}_{22}\in G\ell(N_2;A_g(\mathbf{S}))$. The number of elements of $d^+(\widetilde{B}_{11})$ or of $d^+(\widetilde{B}_{22})$ is less than that of $d^+(A)$ and the result follows by induction. □

REMARK. It is easily seen that theorem 18.13 remains valid if S_o is a sector of the form $S_o=S(-\frac{\pi}{2}+\ell\pi,\frac{\pi}{2}+\ell\pi)$, $\ell\in\mathbb{Z}$.

One would like to have a similar result for the case that S_o is an upper or lower half plane. Apparently, this cannot be achieved by the use of the block diagonalization theorems 17.12 and 17.13 . However, with the aid of theorems 17.17 and 17.18 we can come very close to our goal. To this end, we shall arrange the blocks in the canonical form A^c in a special order.

DEFINITION. *Let* $A\in\hat{M}_o$ *and let* $\alpha=\ell\pi$, $\ell\in\mathbb{Z}$. *A right canonical form of* A *will be denoted by* A^c_α *if the order of the diagonal blocks* $A^c_j(=A^c_{+,j})$, $j=1,..,m$, *is such that, for all pairs* $(i,j)\in\{1,..,m\}\times\{1,..,m\}$ *with* $i>j$, *the following conditions hold*

(i) $(d_i-d_j)e^{i\alpha}\le 0$

(ii) $\mathrm{Re}\{(\mu_{i,k}-\mu_{j,k})e^{ik\alpha}\}\le 0$ *for all* $k\in k(\sigma(A))$ *such that* $k>0$.

The matrix components $A^c_j=A^c_{+,j}$ of A^c_α defined in (2.4), can be grouped in such a way that

(i) the resulting blocks $\widetilde{A}^c_h$ have the form

$$\widetilde{A}^c_h=(s+1)^{d_h(s+1)}s^{-d_hs}\,e^{q_h(s+1)-q_h(s)}(1+\frac{1}{s})^{\widetilde{G}_h},\quad h=1,..,\widetilde{m},\tag{18.14}$$

where d_h and q_h are defined as in (2.4), $\widetilde{G}_h$ is a Jordan matrix and $\widetilde{m}\le m$.

(ii) $h\neq i$ implies that either $d_h\neq d_i$ or $\overline{q_h\neq q_i}$. In this manner, a new partition is formed which, in general, is coarser than the one considered so far. This coarser partition will be denoted as the *-partition.

PROPOSITION 18.15. *Let* $A \in M_g(S_o)$, *where* $S_o = S(\alpha_1^o, \alpha_2^o) = S(-\pi, \pi)$ *or* $S(0, 2\pi)$ *and* $g \in [0,1]$. *Let* $\alpha \in \{\alpha_1^o, \alpha_2^o\}$ *and let* A_α^c *be a right canonical form of* A *as defined above. If*

$$g \leq \min\{k \in k(\sigma(A)) : k > 0\}$$

then there exists a matrix function $F \in G\ell(n; A_g(S_o)[s^{1/p}])$ *such that*

$$F(s+1)^{-1} A(s) F(s) = A_\alpha^c(s) + \tilde{A}(s),$$

where $\tilde{A} \in End(n; A_g(S_o))$ *and* $\tilde{A}$ *is an upper triangular matrix.*

PROOF: As in the preceding theorems, we begin by performing a transformation of the form (18.8). We then distinguish four different situations:

1. Not all d_j are equal $(j = 1,..,m)$.
2. All d_j are equal but not all polynomials $\underline{q_j}$ are equal $(j = 1,..,m)$.
3. All d_j and all $\underline{q_j}$ are equal but not all γ_j are equal $(j = 1,..,m)$.
4. All $d_j, \underline{q_j}$ and γ_j are equal $(j = 1,..,m)$.

In each of the first three cases we can reduce m by application of theorem 17.14. The conclusion of the proposition follows by induction on m.

The details of this proof closely resemble those of the proofs of theorems 18.6 and 18.13 and are therefore omitted. □

Let $A \in \hat{M}_o$, $k \in k(\sigma(A))$ and $\alpha = \ell\pi$, $\ell \in \mathbb{Z}$. Suppose that $\alpha \notin \Sigma_k^+(\sigma(A))$. This implies that $\frac{\pi}{2} - \arg(\mu_{i,k} - \mu_{j,k}) \neq k\alpha \bmod \pi$ for all $i,j \in \{1,..,m\}$, or, equivalently,

$$Re\{(\mu_{i,k} - \mu_{j,k})e^{ik\alpha}\} \neq 0 \quad \text{for all } i,j \in \{1,..,m\}.$$

Application of theorem 17.17 yields the following result (cf. the remark on p.117).

THEOREM 18.16. *Let* $t \in \{1,2\}$ *and let* α_1^o, α_2^o *be real numbers such that*

$$\alpha_t^o = \ell\pi,\ \ell \in \mathbb{Z},\ \text{and}\ \pi < \alpha_2^o - \alpha_1^o \leq 2\pi.$$

Let $S_o = S(\alpha_1^o, \alpha_2^o)$ *and* $A \in M_g(S_t)$, *where* $g \in [0,1]$ *and* S_t *is defined by* (15.11). *Assume that*

(i) $g \leq \min\{k \in k(\sigma(A)) : k > 0\}$
(ii) $\alpha_t^o \notin \Sigma^+(\sigma(A))$.

Then there exists a matrix function $F \in G\ell(n; A_g(S_t)[s^{1/p}])$ *such that*

$$F(s+1)^{-1}A(s)F(s) = A^c_{\alpha^o_t}(s) + s^{-r_o}\tilde{A}(s),$$

where $r_o \in \frac{1}{p}\mathbb{Z}$, $r_o > 1 - \min d^+(A)$, $\tilde{A} \in \mathrm{End}(n;A_g(S_t))$ *and* $\tilde{A}$ *is upper block triangular in the *-partition defined on p.* 122.

That we get a coarser partition here than in proposition 18.15 is due to the fact that theorem 17.17 cannot be used to 'separate' two different elements γ_i and γ_j of $\gamma^+(A)$, if $d_i = d_j$ and $\underline{q_i} = \underline{q_j}$ $(i,j \in \{1,..,m\})$.

<u>THEOREM</u> 18.17. *Let* $A \in M_g(S_t)$, *where* $g \in [0,1]$, $t \in \{1,2\}$ *and* S_t *is defined as in theorem* 18.16. *Let* $\tilde{S}_t = S[\alpha^o_1, \alpha^o_1 + \pi)$ *if* $t = 1$, $\tilde{S}_t = S(\alpha^o_2 - \pi, \alpha^o_2]$ *if* $t = 2$. *Let* A^c *be a right canonical form of* A. *Under the same assumptions as in theorem* 18.16 *there exists a matrix function* $F \in G\ell(n;A_g(\tilde{S}_t)[s^{1/p}])$ *such that*

$$F(s+1)^{-1}A(s)F(s) = A^c(s).$$

<u>PROOF</u>: We shall sketch the proof for the case that $t = 1$ and $\alpha^o_1 = 0$. In view of theorem 18.16 we may assume that A has the form

$$A(s) = A^c_o(s) + s^{-r_o}\tilde{A}(s),$$

where $r_o \in \frac{1}{p}\mathbb{Z}$, $r_o > 1 - \min d^+(A)$, $\tilde{A} \in \mathrm{End}(n;A_g(S_1))$, and $\tilde{A}$ is upper block-triangular in the *-partition defined on p. 122. The assertion of theorem 18.17 can be proved by means of induction on $\tilde{m}$ (cf.(18.14)). If $\tilde{m} = 1$, then F must satisfy an equation of the form

$$F(s+1) - \{(1+s^{-1})^{\tilde{G}} + s^{-1-1/p}C(s)\}F(s)(1+s^{-1})^{-\tilde{G}} = 0,$$

where $\tilde{G}$ is a Jordan matrix and $C \in \mathrm{End}(n;A_g(S_1))$. It follows from proposition 16.21 and theorem 15.15 that this equation possesses a solution $F \in \mathrm{End}(n;A_g(\tilde{S}_1))$ with the property that $F(\infty) = I$, hence $F \in G\ell(n;A_g(\tilde{S}_1))$. Now let $\tilde{m} > 1$. Let us assume that not all d_h are equal $(h = 1,..,\tilde{m})$. Suppose that $d_h = d_1$ for $h = 1,..,\tilde{m}_1$, where $\tilde{m}_1 < \tilde{m}$, and $d_h < d_1$ for $h = \tilde{m}_1 + 1,..,\tilde{m}$. Let $N_1 = \tilde{n}_1 + .. + \tilde{n}_{\tilde{m}_1}$, where $\tilde{n}_h$ is the order of $\tilde{A}^c_h (h = 1,..,\tilde{m})$, and $N_2 = n - N_1$. If we partition A along the N_1 th row and column and put

$$A = s^{d_1}\begin{pmatrix} A_{11} & A_{12} \\ 0 & A_{22} \end{pmatrix},$$

we have $A_{12}(\infty) = A_{22}(\infty) = 0$, whereas $A_{11}(\infty)$ is invertible. According to theorem 17.18 there exists a matrix function $F \in G\ell(n;A_g(\tilde{S}_1))$ of the form

$$\begin{pmatrix} I_{N_1} & F_{12} \\ 0 & I_{N_2} \end{pmatrix} \text{ such that } F(s+1)^{-1}A(s)F(s) = s^{d_1}\operatorname{diag}\{A_{11}(s), A_{22}(s)\} \text{ (cf.(17.5))}.$$

Obviously, both matrix functions A_{11} and A_{22} contain a smaller number of blocks than A.

In the case that all d_h are equal $(h = 1,..,\tilde{m})$, the number $\tilde{m}$ can be reduced in a similar manner. Hence the result follows. □

Now suppose that A and A^{-1} are meromorphic at infinity. Application of theorems 18.13 and 18.17 yields the following result.

THEOREM 18.18 . *Let* $A \in G\ell(n;\mathbb{C}\{s^{-1}\}[s]) \cap \hat{M}_o$ *and let* A^c *be a right canonical form of* A. *Let* $g \in [0,1)$ *and* $g \leq \min\{k \in k(\sigma(A)) : k > 0\}$. *If there is an integer* ℓ *such that*

$$\ell\pi \notin \Sigma^+(\sigma(A)),$$

then there exist matrix functions $F_i \in G\ell(n;A_g(S_i)[s^{1/p}])$, $i = 1,..,4$, *where* $S_1 = S((\ell-\frac{3}{2})\pi, (\ell-\frac{1}{2})\pi)$, $S_2 = S((\ell-1)\pi, \ell\pi]$, $S_3 = S((\ell-\frac{1}{2})\pi, (\ell+\frac{1}{2})\pi)$, $S_4 = S[\ell\pi, (\ell+1)\pi)$, *such that*

$$F_i(s+1)^{-1}A(s)F_i(s) = A^c(s), \qquad i = 1,..,4.$$

Note that the sectors S_1, S_2, S_3 and S_4 cover a neighbourhood of ∞.

HISTORICAL REMARKS

The asymptotic theory of analytic difference equations has its origin in a well-known paper by Poincaré, published in 1885 ([27]). Around the beginning of this century several mathematicians, among whom Birkhoff, Horn, Nörlund and Pincherle, contributed to its development. In the period between the 1930's and the 1960's the theory of difference equations seems to have suffered some neglect and, consequently, it is still in a rather backward state compared to the related field of differential equations. Lately, however, there has been renewed interest in this subject and recent developments in the theory of differential equations have inspired new research on analytic difference equations.

Below we have sketched briefly the evolution of this theory since the beginning of the 20th century. Our account of the work that has been done in this field is far from being complete and the references to particular publications are intended mainly to illustrate different asymptotic methods (for an extensive bibliography of the literature before 1924 we refer the reader to [25]).

One can distinguish between two main approaches: the first uses explicit representations of solutions of difference equations and studies their asymptotic behaviour, whereas the second and more general approach takes formal solutions of these equations as its starting point.

I. *Integral and series representations.*

The following three types of representations are frequently used in asymptotic studies on difference equations.

1. *The Mellin transform.*

Linear difference and linear differential equations with rational coefficients are related (formally) through a Mellin transformation (i.e. a multiplicative Laplace transformation, often simply called Laplace transformation in the earlier literature). This fact may be exploited to represent solutions of linear difference equations by integrals of the form

$$\int_{\gamma} \varphi(u) u^{x-1}\, du,$$

where φ is a solution of the corresponding differential equation and γ a suitable path of integration.

The problem of the existence and the asymptotic behaviour of solutions of this type was studied by Barnes, Galbrun, Nörlund, Pincherle ([26]) and others in the beginning of this century. The method is particularly succesful in the case that the corresponding differential equation is of the Fuchsian type. This situation is often referred to as the 'regular case' (cf. [1] for a different, but equivalent, definition). 'Irregular cases' have been considered by Barnes ([2]), Galbrun ([13]) and, very recently, by Duval ([10]). Unfortunately, the study of these cases is technically very complicated and it appears doubtful whether this method continues to be applicable in the most general case.

Using a slightly different but related approach, Ramis ([32]) has given a very general treatment of the homogeneous linear difference equation with rational coefficients, based on recent results of Malgrange in the theory of differential equations. Duval ([12]) has extended his results to equations whose coefficients admit factorial series representations in a right or a left half plane.

2. *Laplace integrals.*

Another powerful tool in the asymptotic theory of analytic difference equations is the Laplace transformation. A (formal) inverse Laplace transformation carries a linear or nonlinear difference equation into an integral equation. Unlike the preceding method, this technique is not restricted to equations with rational coefficients and it yields, in principle, solutions with the desired asymptotic properties in half planes with almost arbitrary orientation (solutions obtained with the aid of the Mellin transformation usually have the required asymptotic behaviour in either a left or a right half plane). In other respects, however, its domain of applicability is more limited than that of the Mellin transform.

In [18] Horn treats the 'regular case' of the homogeneous linear difference equation with the aid of Laplace transform techniques. A detailed analysis of the second order homogeneous linear difference equation can be found in a paper by Culmer and Harris ([9], cf. also [7]).

Solutions of nonlinear difference equations in the form of Laplace integrals have been studied by Horn, in [19], and by Harris and Sibuya in [17]. The most general results in this direction were obtained by Braaksma in [6].

3. *Factorial series.*

From the algebraic point of view, the role played by factorial series in the theory of difference equations is similar to that of power series in the theory of differential equations.

Convergent factorial series representations of solutions of linear difference equations have been studied extensively by Nörlund ([23],[24]) and, more recently, by Harris ([14],[15]). These series are closely related to the theory of the Laplace transformation and are frequently considered in combination with Laplace integral representations (cf. [6],[17],[38]).

II. *Existence theorems*.

In situations where no integral or series representations of solutions are available, it is often still possible to establish the existence of these solutions by a different method. Once the corresponding formal problem has been solved, it remains to be shown that the formal solutions of the equation are asymptotic to actual, i.e. analytic solutions.

In some special cases the application of a fixed point theorem or iteration techniques may lead directly to the desired result (cf. [16],[35]). However, in general the procedure will consist of several steps. First, a class of equations of a particularly simple type (usually first order linear equations) are studied. The results of this analysis are then extended to more general classes of equations by means of perturbation or other arguments.

Methods of this type were developed by Birkhoff ([4],[5]), Adams ([1]) and Carmichael ([8]). In [5] Birkhoff and Trjitzinsky claim to have given a complete analytic theory of the homogeneous linear system of difference equations. However, some of the proofs are incomplete and not all the conclusions seem to be justified. Batchelder has written an instructive book on linear difference equations ([3]), in which he presents a systematic treatment of the homogeneous linear difference equation of the 2nd order with linear coefficients. This book contains a clear exposition of Birkhoff's (earlier) ideas as well as a detailed discussion of integral and series representations of solutions.

The approach we have chosen in this monograph belongs to the second category. It is inspired by such methods in the theory of differential equations as have been developed by Hukuhara, Sibuya, Malgrange and others (cf. [20],[21],[34], [40]).

REFERENCES

[1] ADAMS, C.R., *On the irregular cases of the linear ordinary difference equation*, Trans.Amer.Soc., 30 (1928), 507-541.

[2] BARNES, E.W., *On the homogeneous linear difference equation of the second order with linear coefficients*, Messenger of Math., 34 (1905), 52-71.

[3] BATCHELDER, P.M., *An introduction to linear difference equations*, Dover Publications, New York (1967).

[4] BIRKHOFF, G.D., *General theory of linear difference equations*, Trans.Amer.Math.Soc., 12 (1911), 243-284.

[5] BIRKHOFF, G.D. and W.J. TRJITZINSKY, *Analytic theory of singular difference equations*, Acta Math., 60 (1933), 1-89.

[6] BRAAKSMA, B.L.J., *Laplace integrals in singular differential and difference equations*, Ordinary and Partial Diff.Eq., Proc. Dundee 1978, Lecture Notes in Mathematics, 827, 25-53, Springer Verlag, 1980.

[7] BRAAKSMA B.L.J. and W.A. HARRIS Jr., *On an open problem in the theory of linear difference equations*, Nieuw Archief voor Wiskunde (3), 23 (1975), 228-240.

[8] CARMICHAEL, R.D., *On the solutions of linear homogeneous difference equations*, Amer. J. Math., 38 (1916), 185-220.

[9] CULMER, W.J.A. and W.A. HARRIS Jr. *Convergent solutions of ordinary linear homogeneous difference equations*, Pacific J. Math., 13 (1968), 1111-1138.

[10] DUVAL, A., *Solutions irrégulières d'équations aux différences polynomiales*, and *Equations aux différences algébriques: solutions méromorphes dans* $\mathbb{C}$, Equations différentielles et systèmes de Pfaff dans le plan complexe II, Lecture Notes in Mathematics, 1015, 64-135, Springer Verlag, 1983.

[11] DUVAL, A., *Lemmes de Hensel et factorisation formelle pour les opérateurs aux différences*, to appear in Funkcialaj Ekvacioj, 26 (1984).

[12] DUVAL, A., *Equations aux différences dans le champ complexe*, thesis, Publ. I.R.M.A. Strasbourg (1984).

[13] GALBRUN, H., *Sur certaines solutions exceptionnelles d'une équation linéaire aux différences finies*, Bull. S.M.F., 49 (1921), 206-241.

[14] HARRIS, W.A. Jr., *Linear systems of difference equations*, Contributions to differential equations, 1 (1963), 489-518.

[15] HARRIS, W.A. Jr. *Equivalent classes of difference equations*, Contributions to differential equations, 2 (1963), 253-264.

[16] HARRIS, W.A., Jr. and Y. SIBUYA, *Asymptotic solutions of systems of nonlinear difference equations*, Arch. Rat. Mech. Anal., 15 (1964), 377-395.

[17] HARRIS, W.A. Jr. and Y. SIBUYA, *On asymptotic solutions of systems of nonlinear difference equations*, J. reine angew. Math., 222 (1966), 120-135.

[18] HORN, J., *Zur Theorie der linearen Differenzengleichungen*, Jahresb. deutsch. Math. Ver., 24 (1915), 210-225.

[19] HORN, J., *Über eine nichtlineare Differenzengleichung*, Jahresb. deutsch. Math. Ver., 26 (1918), 230-251.

[20] HUKUHARA, M., *Sur les points singuliers des équations différentielles linéaires* II, J. Fac. Sci. Hokkaido Imp. Univ. 5 (1937), 123-166, and III, Mém. Fac. Sci. Kyushu Univ., Ser. A, 2 (1941), 125-137.

[21] MALGRANGE, B., *Sur les points singuliers des équations différentielles*, l'Enseign. Math., 20, 1-2 (1974), 147-176.

[22] NEVANLINNA, F., *Zur Theorie der Asymptotischen Potenzreihen*, Annales Academiae Scientiarum Fennicae, Ser. A, 12 (1919).

[23] NÖRLUND, N.E., *Sur l'intégration des équations linéaires aux différences finies par séries de facultés*, Rend. Circ. Mat. Palermo, 35 (1913), 177-216.

[24] NÖRLUND, N.E., *Leçons sur les séries d'interpolation*, Gauthiers-Villars, Paris (1926).

[25] NÖRLUND, N.E., *Vorlesungen über Differenzenrechnung*, Chelsea, New York (1954).

[26] PINCHERLE, S., *Sur la génération des systèmes récurrents au moyen d'une équation linéaire différentielle*, Acta Math. 16 (1892), 341-363.

[27] POINCARE, H., *Sur les équations linéaires aux différentielles ordinaires et aux différences finies*, Am. J. of Math. 7 (1885), 203-258.

[28] PRAAGMAN, C., *The formal classification of linear difference operators*, Proceedings Kon. Nederl. Ac. van Wetensch., ser. A., 86 (2) (1983), 249-261.

[29] RAMIS, J.P., *Dévissage Gevrey*, Astérisque S.M.F., 59-60 (1978), 173-204.

[30] RAMIS, J.P., *Les séries k-sommables et leurs applications*, Springer Lecture Notes in Physics, 126 (1980), 178-199.

[31] RAMIS, J.P., *Théorèmes d'indices Gevrey pour les équations différentielles*, Mem. Amer. Math. Soc., 296 (1984).

[32] RAMIS, J.P., *Etude des solutions méromorphes des équations aux différences linéaires algébriques*, to appear.

[33] SCHECHTER, M., *Principles of Functional Analysis*, Academic Press, New York (1971).

[34] SIBUYA, Y., *Simplification of a system of linear ordinary differential equations about a singular point*, Funkcialaj Ekvacioj, 4 (1962), 29-56.

[35] TANAKA, S., *On asymptotic solutions of nonlinear difference equations* I, II, III, Mem. Fac. Sci. Kyushu Univ., Ser. A, 7 (1953), 107-127, 10 (1956), 45-83, 11 (1957), 167-184.

[36] TITCHMARSH, E.C., *The Theory of Functions*, (2nd ed.), Oxford University Press, Oxford (1939).

[37] TRJITZINSKY, W.J., *Nonlinear difference equations*, Compositio Math., 5 (1938), 1-60.

[38] TRJITZINSKY, W.J., *Laplace integrals and factorial series in the theory of linear differential and difference equations*, Trans. Amer. Math. Soc., 37 (1935), 80-146.

[39] TURRITTIN, H.L., *The formal theory of systems of irregular homogeneous linear difference and differential equations*, Bol. Soc. Math. Mexicana (1960), 255-264.

[40] WASOW, W., *Asymptotic Expansions for Ordinary Differential Equations*, Interscience Publishers, New York (1965).

LIST OF SYMBOLS AND NOTATIONS

We use the following notations:

$k[x]$: the set of polynomials in x with coefficients in k.

$k[[x]]$: the set of formal power series in x with coefficients in k.

$k\{x\}$: the set of convergent power series in x with coefficients in k.

$End(n;X)$: the set of linear mappings from X^n into itself.

$G\ell(n;X)$: the set of invertible linear mappings from X^n onto itself.

List of frequently occurring symbols:

Symbol	p.	Symbol	p.
A_+^c	5	$R_N(f;s)$	10
A_-^c	9	$S(\alpha_1,\alpha_2)$	12
$d^+(A)$	8	$S[\alpha_1,\alpha_2)$	12
$\gamma^+(A)$	8	$S(\alpha_1,\alpha_2]$	12
$k^+(A)$ $(=k(A))$	8	$S[\alpha_1,\alpha_2]$	12
$\mu_k^+(A)$	8	$\overline{S(\alpha_1,\alpha_2)}$	12
$\Sigma^+_{1,j}(A)$	8	$S_R(\alpha_1,\alpha_2)$	11
$\Sigma^+_k(A)$	8	$B_{g,a,r}(G)$	14
$\Sigma^+(A)$	8	$R_N(\varphi;s,y)$	77
$\underline{q}$	7	$V_o(F,y_o)$	77
degr	8	$V_g(F,y_o)$	78
$\sum_{ph=o}^{p}$	6	$b_{g,\alpha,\rho}$	92
$\hat{M}_o$	7	$\tau_\pm$	93
$\hat{M}_g^o$	11	$\delta_\pm$	93
$A_g(F)$	10	$\chi(L;X,Y)$	96
$A_{go}(F)$	10	$B(X,Y)$	96
$M_g(F)$	11	$\hat{V}^{(2)}_{g,\alpha,\rho}(y_o)$	101
$¢([[s^{-1/p}]]_{pg})$	11	$\hat{V}^{(2)}_g(y_o)$	101
		$\sigma(A)$	118

INDEX

Vol. 981: Value Distribution Theory. Proceedings, 1981. Edited by I. Laine and S. Rickman. VIII, 245 pages. 1983.

Vol. 982: Stability Problems for Stochastic Models. Proceedings, 1982. Edited by V. V. Kalashnikov and V. M. Zolotarev. XVII, 295 pages. 1983.

Vol. 983: Nonstandard Analysis-Recent Developments. Edited by A. E. Hurd. V, 213 pages. 1983.

Vol. 984: A. Bove, J. E. Lewis, C. Parenti, Propagation of Singularities for Fuchsian Operators. IV, 161 pages. 1983.

Vol. 985: Asymptotic Analysis II. Edited by F. Verhulst. VI, 497 pages. 1983.

Vol. 986: Séminaire de Probabilités XVII 1981/82. Proceedings. Edited by J. Azéma and M. Yor. V, 512 pages. 1983.

Vol. 987: C. J. Bushnell, A. Fröhlich, Gauss Sums and p-adic Division Algebras. XI, 187 pages. 1983.

Vol. 988: J. Schwermer, Kohomologie arithmetisch definierter Gruppen und Eisensteinreihen. III, 170 pages. 1983.

Vol. 989: A. B. Mingarelli, Volterra-Stieltjes Integral Equations and Generalized Ordinary Differential Expressions. XIV, 318 pages. 1983.

Vol. 990: Probability in Banach Spaces IV. Proceedings, 1982. Edited by A. Beck and K. Jacobs. V, 234 pages. 1983.

Vol. 991: Banach Space Theory and its Applications. Proceedings, 1981. Edited by A. Pietsch, N. Popa and I. Singer. X, 302 pages. 1983.

Vol. 992: Harmonic Analysis, Proceedings, 1982. Edited by G. Mauceri, F. Ricci and G. Weiss. X, 449 pages. 1983.

Vol. 993: R. D. Bourgin, Geometric Aspects of Convex Sets with the Radon-Nikodým Property. XII, 474 pages. 1983.

Vol. 994: J.-L. Journé, Calderón-Zygmund Operators, Pseudo-Differential Operators and the Cauchy Integral of Calderón. VI, 129 pages. 1983.

Vol. 995: Banach Spaces, Harmonic Analysis, and Probability Theory. Proceedings, 1980–1981. Edited by R.C. Blei and S.J. Sidney. V, 173 pages. 1983.

Vol. 996: Invariant Theory. Proceedings, 1982. Edited by F. Gherardelli. V, 159 pages. 1983.

Vol. 997: Algebraic Geometry – Open Problems. Edited by C. Ciliberto, F. Ghione and F. Orecchia. VIII, 411 pages. 1983.

Vol. 998: Recent Developments in the Algebraic, Analytical, and Topological Theory of Semigroups. Proceedings, 1981. Edited by K. H. Hofmann, H. Jürgensen and H. J. Weinert. VI, 486 pages. 1983.

Vol. 999: C. Preston, Iterates of Maps on an Interval. VII, 205 pages. 1983.

Vol. 1000: H. Hopf, Differential Geometry in the Large, VII, 184 pages. 1983.

Vol. 1001: D.A. Hejhal, The Selberg Trace Formula for PSL(2, IR). Volume 2. VIII, 806 pages. 1983.

Vol. 1002: A. Edrei, E.B. Saff, R.S. Varga, Zeros of Sections of Power Series. VIII, 115 pages. 1983.

Vol. 1003: J. Schmets, Spaces of Vector-Valued Continuous Functions. VI, 117 pages. 1983.

Vol. 1004: Universal Algebra and Lattice Theory. Proceedings, 1982. Edited by R. S. Freese and O. C. Garcia. VI, 308 pages. 1983.

Vol. 1005: Numerical Methods. Proceedings, 1982. Edited by V. Pereyra and A. Reinoza. V, 296 pages. 1983.

Vol. 1006: Abelian Group Theory. Proceedings, 1982/83. Edited by R. Göbel, L. Lady and A. Mader. XVI, 771 pages. 1983.

Vol. 1007: Geometric Dynamics. Proceedings, 1981. Edited by J. Palis Jr. IX, 827 pages. 1983.

Vol. 1008: Algebraic Geometry. Proceedings, 1981. Edited by J. Dolgachev. V, 138 pages. 1983.

Vol. 1009: T. A. Chapman, Controlled Simple Homotopy Theor Applications. III, 94 pages. 1983.

Vol. 1010: J.-E. Dies, Chaînes de Markov sur les permutation 226 pages. 1983.

Vol. 1011: J. M. Sigal. Scattering Theory for Many-Body Qua Mechanical Systems. IV, 132 pages. 1983.

Vol. 1012: S. Kantorovitz, Spectral Theory of Banach Space ators. V, 179 pages. 1983.

Vol. 1013: Complex Analysis – Fifth Romanian-Finnish Ser Part 1. Proceedings, 1981. Edited by C. Andreian Cazacu, N. B M. Jurchescu and I. Suciu. XX, 393 pages. 1983.

Vol. 1014: Complex Analysis – Fifth Romanian-Finnish Sen Part 2. Proceedings, 1981. Edited by C. Andreian Cazacu, N. B M. Jurchescu and I. Suciu. XX, 334 pages. 1983.

Vol. 1015: Equations différentielles et systèmes de Pfaff da champ complexe – II. Seminar. Edited by R. Gérard et J. P. R V, 411 pages. 1983.

Vol. 1016: Algebraic Geometry. Proceedings, 1982. Edited b Raynaud and T. Shioda. VIII, 528 pages. 1983.

Vol. 1017: Equadiff 82. Proceedings, 1982. Edited by H. W. Kno and K. Schmitt. XXIII, 666 pages. 1983.

Vol. 1018: Graph Theory, Łagów 1981. Proceedings, 1981. Edit M. Borowiecki, J. W. Kennedy and M. M. Sysło. X, 289 pages.

Vol. 1019: Cabal Seminar 79–81. Proceedings, 1979–81. Edit A. S. Kechris, D. A. Martin and Y. N. Moschovakis. V, 284 pages.

Vol. 1020: Non Commutative Harmonic Analysis and Lie Gr Proceedings, 1982. Edited by J. Carmona and M. Vergne. V pages. 1983.

Vol. 1021: Probability Theory and Mathematical Statistics. ceedings, 1982. Edited by K. Itô and J.V. Prokhorov. VIII, 747 p 1983.

Vol. 1022: G. Gentili, S. Salamon and J.-P. Vigué. Geometry Se "Luigi Bianchi", 1982. Edited by E. Vesentini. VI, 177 pages. 19

Vol. 1023: S. McAdam, Asymptotic Prime Divisors. IX, 118 p 1983.

Vol. 1024: Lie Group Representations I. Proceedings, 1982– Edited by R. Herb, R. Lipsman and J. Rosenberg. IX, 369 pages.

Vol. 1025: D. Tanré, Homotopie Rationnelle: Modèles de Quillen, Sullivan. X, 211 pages. 1983.

Vol. 1026: W. Plesken, Group Rings of Finite Groups Over Integers. V, 151 pages. 1983.

Vol. 1027: M. Hasumi, Hardy Classes on Infinitely Connecte mann Surfaces. XII, 280 pages. 1983.

Vol. 1028: Séminaire d'Analyse P. Lelong – P. Dolbeault – H. S Années 1981/1983. Edité par P. Lelong, P. Dolbeault et H. S VIII, 328 pages. 1983.

Vol. 1029: Séminaire d'Algèbre Paul Dubreil et Marie-Paule Mal Proceedings, 1982. Edité par M.-P. Malliavin. V, 339 pages. 19

Vol. 1030: U. Christian, Selberg's Zeta-, L-, and Eisensteins XII, 196 pages. 1983.

Vol. 1031: Dynamics and Processes. Proceedings, 1981. Edit Ph. Blanchard and L. Streit. IX, 213 pages. 1983.

Vol. 1032: Ordinary Differential Equations and Operators. ceedings, 1982. Edited by W. N. Everitt and R. T. Lewis. XV, 521 p 1983.

Vol. 1033: Measure Theory and its Applications. Proceedings, Edited by J. M. Belley, J. Dubois and P. Morales. XV, 317 pages.

Vol. 1034: J. Musielak, Orlicz Spaces and Modular Spac 222 pages. 1983.